ENERGY-LINKED FUNCTIONS
OF
MITOCHONDRIA

ENERGY-LINKED FUNCTIONS

OF

MITOCHONDRIA

Edited by
Britton Chance
Johnson Research Foundation, University of Pennsylvania
Philadelphia, Pennsylvania

Papers presented at the First Colloquium of the
Johnson Research Foundation of the University of Pennsylvania
Philadelphia, April 13, 1963

1963

 ACADEMIC PRESS New York • London

Copyright © 1963 by Academic Press Inc.

ALL RIGHTS RESERVED

NO PART OF THIS BOOK MAY BE REPRODUCED IN ANY FORM
BY PHOTOSTAT, MICROFILM, OR ANY OTHER MEANS,
WITHOUT WRITTEN PERMISSION FROM THE PUBLISHERS.

ACADEMIC PRESS INC.
111 Fifth Avenue
New York 3, N. Y.

United Kingdom Edition
Published by
ACADEMIC PRESS INC. (London) Ltd.
Berkeley Square House, London W. 1

Library of Congress Catalog Card Number: *63-20025*

PRINTED IN THE UNITED STATES OF AMERICA

CONTRIBUTORS

Alm, B., The Wenner-Gren Institute, University of Stockholm, Stockholm, Sweden, p. 5

Andreoli, T. E., Gerontology Branch, National Heart Institute, National Institutes of Health, PHS, U.S. Department of Health, Education & Welfare, Bethesda, and the Baltimore City Hospitals, Baltimore, Maryland, p. 26

Bose, S. K., The Henry Shaw School of Botany and The Adolphus Busch III Laboratory of Molecular Biology, Washington University, St. Louis, Missouri, p. 207

Brierley, G. P., Institute for Enzyme Research, University of Wisconsin, Madison, Wisconsin, p. 237

Chance, B., Johnson Research Foundation, University of Pennsylvania, Philadelphia, Pennsylvania, p. 253

Chappell, J. B., The Department of Biochemistry, University of Cambridge, Cambridge, England, p. 219

Cohn, M., Johnson Research Foundation, University of Pennsylvania, Philadelphia, Pennsylvania, p. 217

Danielson, L., The Wenner-Gren Institute, University of Stockholm, Stockholm, Sweden, p. 157

Ernster, L., The Wenner-Gren Institute, University of Stockholm, Stockholm, Sweden, p. 157

Estabrook, R. W., Johnson Research Foundation, University of Pennsylvania, Philadelphia, Pennsylvania, pp. 143

Gest, H., The Henry Shaw School of Botany and The Adolphus Busch III Laboratory of Molecular Biology, Washington University, St. Louis, Missouri, p. 207

Gonze, J., Johnson Research Foundation, University of Pennsylvania, Philadelphia, Pennsylvania, p. 143

Greville, G. D., The Biochemistry Department, Institute of Animal Physiology, Babraham, Cambridge, England, p. 219

Hommes, F. A., Johnson Research Foundation, University of Pennsylvania, Philadelphia, Pennsylvania, p. 39 and p. 143

Klingenberg, M., Physiologisch-Chemisches Institut der Universität Marburg, Marburg, Germany, p. 121

Löw, H., The Wenner-Gren Institute, University of Stockholm, Stockholm, Sweden, p. 5

Marchant, R. H., Department of Physiology, University of California, Berkeley, California, p. 51

Mukohata, Y., Department of Physiology, University of California, Berkeley, California, p. 51

Packer, L., Department of Physiology, University of California, Berkeley, California, p. 51

Penefsky, H. S., The Public Health Research Institute of The City of New York, Inc., New York, New York, p. 87

Pharo, R. L., Gerontology Branch, National Heart Institute, National Institutes of Health, PHS, U.S. Department of Health Education & Welfare, Bethesda, and the Baltimore City Hospitals, Baltimore, Maryland, p. 26

Pressman, B. C., Johnson Research Foundation, University of Pennsylvania, Philadelphia, Pennsylvania, p. 181

Racker, E., The Public Health Research Institute of The City of New York, Inc., New York, New York, p. 75

Sanadi, R., Gerontology Branch, National Heart Institute, National Institutes of Health, PHS, U. S. Department of Health, Education & Welfare, Bethesda, and the Baltimore City Hospitals, Baltimore, Maryland, p. 26

Slater, E. C., Laboratory of Physiological Chemistry, University of Amsterdam, The Netherlands, p. 97

Tager, J. M., Laboratory of Physiological Chemistry, University of Amsterdam, The Netherlands, p. 97

Vallin, J., The Wenner-Gren Institute, University of Stockholm, Stockholm, Sweden, p. 5

Vyas, S. R., Gerontology Branch, National Heart Institute, National Institutes of Health, PHS, U.S. Department of Health, Education & Welfare, Bethesda, and the Baltimore City Hospitals, Baltimore, Maryland, p. 26

PREFACE

There occur rare occasions when the scientific topic, the group of experts, and a convenient place come together--perhaps fortuitously--to give rise nearly spontaneously to an exchange of ideas and a community of thought leading to the emergence of concepts--which surely may have preexisted--but which are brought clearly into focus. It is possible that exactly this may have happened here.

This colloquium was loosely organized and initially had as its primary object an opportunity for a "get-together" of those scientists interested in energy-linked processes in mitochondria in general and reversal of electron transport in particular. Our group is largely composed of those who are passing through Philadelphia on their way to the Federation Meetings, or of those who have stopped at the Johnson Research Foundation for a brief interval of study on a common problem before the Meetings.

As our program developed, and we became more encouraged by the possibility of foreign attendance, the interests of the participants allowed the program to develop along the three general lines of research which constitute the three main divisions of this volume: the use of mitochondrial energy sources for the formation of reduced substances, particularly DPNH and TPNH; the transfer of this reducing power from DPNH to TPNH in transhydrogenase reactions, or in substrate reduction such as glutamate synthesis; and finally, the utilization of such reducing power in ion accumulation or in bacterial photosynthesis.

The key point that has emerged with greater force and clarity from these discussions than was possible in the independent--and excellent--contributions of the participants prior to this meeting is a change in the concept of the role of ATP as the sole energetic reaction product of the mitochondrial reactions. While a good deal of evidence for high-energy precursors of mitochondria had been accumulated, much of it by the participants of this colloquium, the closing ring of evidence for the function of these high-energy intermediates not only in the energy-linked reduction of DPN and TPN, but also in the transport of ions across membranes, now leads to a new and important concept in mitochondrial energy relations, and focuses our attention on a new class of non-phosphorylated high-energy compounds.

We regard this volume to be an experiment in scientific communication--an area in which experiments are more needed than theories. We have emphasized speed in the publication of this colloquium and believe this to be essential in this rapidly developing field. Although we may have lost some of the accuracy, formality, and consistency of a more pretentious volume, it is our hope that rapid availability of viewpoints and experimental data in this field compensate for errors that may have been passed over, and for which the Editor accepts his responsibilities.

The rapid production of this volume owes much to the excellent cooperation of the participants and to the speedy and effective scheduling of Academic Press. But the real work of organization and of gathering and proofing the material has been carried out by Lilian S. L. Chance and Drs. R. W. Estabrook and T. Conover. We are greatly indebted to Elaine S. Polin and Jan Bright for the typing and to Kenneth Ray and Carlos Whiting for the artwork.

Britton Chance

May 28, 1963

CONTENTS

List of Contributors v

Preface vii

Introductory Remarks 1

THE GENERATION OF REDUCING POWER

Some Aspects of Oxidative Phosphorylation and its
Reversal in Submitochondrial Particles 5
H. Löw, T. Vallin and B. Alm

Discussion 17

Reductive Dephosphorylation of ATP in the DPNH-
Coenzyme Q Segment of the Respiratory Chain 26
D.R. Sanadi, T.E. Andreoli, R.E. Pharo, and
S.R. Vyas

Discussion 36

The Role of Pyridine Nucleotide in Energy-linked
Reactions 39
F.A. Hommes

Discussion 49

Coupling of Energy-linked Functions in Mitochondria
and Chloroplasts to the Control of Membrane
Structure 51
L. Packer, R.H. Marchant and Y. Mukohata

Discussion 72

Topography of Coupling Factors in Oxidative
Phosphorylation 75
E. Racker

Discussion 82

The Energy-linked Reduction of Ubiquinone in Beef
Heart Mitochondria 87
H.S. Penefsky

Discussion 92

ENERGY-LINKED FUNCTIONS OF MITOCHONDRIA

THE TRANSFER OF REDUCING POWER

Provision of Reducing Power for Glutamate Synthesis 97
E.C. Slater and J.M. Tager

Discussion 114

Morphological and Functional Aspects of
Pyridine Nucleotide Reactions in Mitochondria. . 121
M. Klingenberg

Discussion 140

The Interaction of Mitochondrial Pyridine
Nucleotides 143
R.W. Estabrook, F. Hommes and J. Gonze

Discussion 153

Energy-Dependent Reduction of TPN by DPNH 157
L. Danielson and L. Ernster

Discussion 176

Specific Inhibitors of Energy Transfer 181
B.C. Pressman

Discussion 200

THE UTILIZATION OF REDUCING POWER

Relationships between Energy-Generation and Net
Electron Transfer in Bacterial Photosynthesis. . 207
S.K. Bose and H. Gest

The Accumulation of Divalent Ions by Isolated
Mitochondria 219
J.B. Chappell, M. Cohn and G.D. Greville

Discussion 232

Ion Accumulation in Heart Mitochondria 237
G.P. Brierley

Discussion 246

Calcium Stimulated Respiration in Mitochondria .. 253
  B. Chance

  Discussion....................... 270

Envoi............................ 275

Index of Participants................... 277

Index to Subjects..................... 279

INTRODUCTORY REMARKS

Britton Chance
Johnson Research Foundation, University of Pennsylvania
Philadelphia, Pennsylvania

It is a great pleasure to welcome you at this minor prelude to the much larger meetings to be held in Atlantic City next week. The small colloquium which we planned in honor of Professor Slater's visit has grown like Topsy, in a very nice way: with very little urging, our good friends and colleagues from laboratories here and abroad have come, often from great distances at considerable inconvenience, for the sake of some good scientific conversation. We are very proud that meetings of this sort can occur under congenial circumstances and without a high degree of organization, and that we may here take up in some detail a topic which cannot be handled adequately at the Federation Meetings.

To my mind, one of the objects of this colloquium is to give particular attention to the work of those in laboratories abroad--particularly in Amsterdam, Marburg, Oxford, Cambridge, and Stockholm, where many important studies and discoveries in the area of energy-linked functions of mitochondria have occurred. Therefore, the colloquium has three general topics: the generation, the transfer, and the utilization of reducing power.

There are many steps involved in the combined electron and energy transfer reactions which lead to the generation of high-energy intermediates, and at least two sites at which energy can be conserved and then used in the most readily observable reaction of these intermediates in the mitochondria --the reduction of endogenous or added DPN. We are fortunate in having with us today six active workers in the field, who will describe their recent studies.

More recently, particularly in Amsterdam and Stockholm, the transfer of energy-linked reducing power to various acceptors has become a topic of current interest. Two types of transfer have been studied, one to hydrogen acceptors such as ketoglutarate and ammonia or acetoacetate, and the other, most recently, to TPN itself, which appears also to be involved with an energy-linked reaction.

ENERGY-LINKED FUNCTIONS OF MITOCHONDRIA

Lastly, ion transport itself can be activated not only by ATP but also by high-energy intermediates of oxidative phosphorylation. The energy demands imposed upon mitochondria by the transport of divalent cations are now being studied in Baltimore, Madison, and Cambridge, and we are happy to have representatives from two of these laboratories here.

I do not think that it is necessary for me to give a lecture or a demonstration of the existence of the energy-linked pathway leading from succinate, or indeed from respiratory carriers such as cytochrome c at a high oxidation-reduction potential, which somehow interacts with the respiratory chain to produce potentials considerably more negative than 320 mV. I might say, however, that it is pleasant to observe that the enigma which confronted us some years ago, when succinate at a high oxidation-reduction potential proved to be the most effective member of the citric acid cycle for reducing DPN, appears now to be on the verge of solution, as judged by the papers of this colloquium. It is most gratifying to note that, instead of being a "dead-end" reaction of little physiological importance or significance, studies of the key role of this reaction in the formation of reducing power in the mitochondria give _in vitro_ evidence of its important physiological function. Lastly, the increasing evidence of the essential role of high-energy, non-phosphorylated precursors of ATP in the energy-linked functions of mitochondria in DPN and TPN reduction and in ion accumulation by mitochondria opens new vistas in the nature and function of energy-linked reactions.

I am gratified that the attendance has been so excellent, but must note that Dr. Lars Ernster was unfortunately unable to be present, although we welcome members of his laboratory in Stockholm. In addition, Dr. Lehninger sent his regrets this morning on being unable to attend.

This is all I have to say with respect to opening remarks; let us proceed with the first paper.

THE GENERATION OF REDUCING POWER

SOME ASPECTS OF OXIDATIVE PHOSPHORYLATION AND ITS REVERSAL IN SUBMITOCHONDRIAL PARTICLES

Hans Löw, Ivar Vallin, and Barbro Alm

The Wenner-Gren Institute, University of Stockholm
Stockholm, Sweden

Considerable amounts of added DPN can be reduced at a high rate by succinate in an anaerobic system, with energy added in the form of ATP (1, 2). This finding offered a good possibility of studying the interaction between the electron-transporting and the energy-utilizing parts of the oxidative phosphorylation complex, and, eventually of studying this interaction at a single one of the three sites of oxidative phosphorylation.

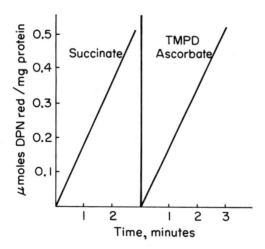

Fig. 1. <u>ATP dependent DPN reduction</u>. The reduction of DPN was followed as an increase of absorption at 340 mµ in a Beckman DK-2 spectrophotometer. 0.45 mg of particle protein was preincubated for 7 minutes in a medium containing 50 mM Tris-HCl pH= 8.0; 6 mM MgCl ; 0.25 M sucrose, 1 mM KCN and 1 mM ATP. The reaction was initiated by the addition of 0.3 mM TMPD; 5 mM ascorbate and 1.5 mM DNP or by 10 mM succinate and 1.5 mM DNP. The final volume was 3.0 ml and the reaction was studied at 30°C.

The observation that the reduction of DPN also took place when other sources of electrons were used, i.e., tetramethylparaphenylene diamine (TMPD) and ascorbate, feeding the electrons into the electron-transport chain at the level of cytochrome c, made possible the comparison of the two system with a purpose to obtain useful information about the mechanism of the second phosphorylation site. As seen in Fig. 1 the rate of reduction of DPN is rather high in both systems, being in this experiment between 0.2 and 0.15 μmoles/min./mg. protein. Generally the succinate system is the more stable and works at a higher rate than the TMPD-ascorbate system.

TABLE I

The effect of antimycin A on the energy dependent reduction of DPN

Experimental conditions as in Fig. 1. Antimycin A was added as 1 μl of an alcoholic solution.

Source of electrons	antimycin A ug/mg protein	μmoles DPN red/ min/mg protein	% inhibition
Succinate	-	0.20	-
	2	0.18	8.5
	4	0.14	29.6
	8	0.07	62.5
	12	0.06	68.4
TMPD and Ascorbate	-	0.18	-
	0.15	0.16	10.5
	0.16	0.14	21
	0.17	0.08	54
	0.18	0.04	76
	0.19	-	100

TMPD = tetramethyl-p-phenylendiazine

Inhibitors such as oligomycin, DNP, rotenone and Amytal inhibit the reduction of DPN according to the same pattern (2, 3), irrespective of the electron donor. As seen from Table 1 the two systems differ in one respect: antimycin A inhibits the TMPD at concentrations many times lower than those needed to produce any effect on the succinate system. This is inherent with the concept introduced earlier (2) that the ascorbate-TMPD system primarily involves the step sensitive to antimycin A whereas the succinate system does not.

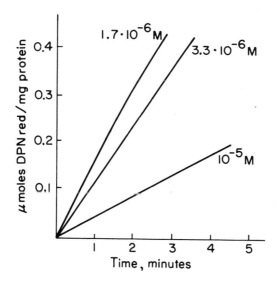

Fig. 2 <u>Phenazine</u> <u>methosulfate</u> <u>as</u> <u>electron</u> <u>mediator</u> <u>in</u> <u>DPN</u> <u>reduction</u>. Preincubation maintained as in Fig. 1. The reaction was started by the addition of the amounts of PMS indicated in the figure. 5 mM ascorbate and 1.5 mM DPN.

The energy-dependent reduction of DPN can be mediated by dyes other than TMPD and also be various quinones. Apart from TMPD, phenzaine methosulphate (PMS) is so far the most efficient mediator of the reduction of DPN by ascorbate. As seen in Fig. 2, PMS in concentrations of 1.7×10^{-6} M gives reduction rates just slightly lower than those maximally obtained with TMPD. This PMS-linked reduction is sensitive to antimycin A and rotenone. It has not been found possible to use cytochrome c to mediate the reduction of DPN.

In these experiments with the reversal of electron transport we have introduced a routing preincubation with ATP prior to the addition of the oxidation-reduction components.

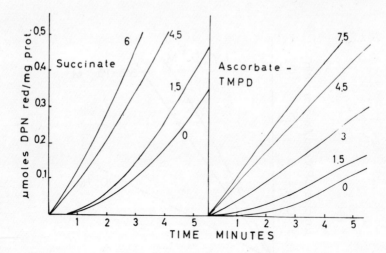

Fig. 3. The effect of preincubation with ATP. Rate of μmoles DPN reduced/mg protein with time. The various preincubation periods in minutes indicated in the figure. Experimental conditions as in Fig. 1.

Fig. 3 shows the effect of preincubation with 1 mM ATP, and it can be seen that in the succinate system the preincubation takes away the lag period, but does not essentially result in any increase in the rate of reduction. When TMPD-ascorbate was the reductant the lag-phase is abolished, but there is also an increase in the rate of reduction. The concentration of ATP needed to achieve maximal effect of the preincubation is indeed much lower than 1 mM. As seen in Fig. 4, preincubation with 10^{-5} M ATP for 7 minutes abolishes the lag phase and brings the rate of reduction almost to the maximum. 2×10^{-5} M ATP is completely effective. These low concentrations of ATP are of course insufficient to drive the reduction. This figure also demonstrates that a preincubation with ATP makes possible the use of ITP to drive the reduction. I will later return to this phenomenon.

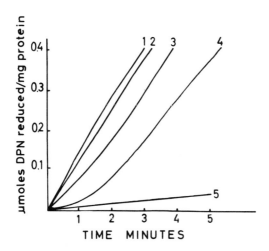

Fig. 4. <u>Reduction of DPN from the cytochrome c level</u>. <u>Preincubation with small amounts of ATP</u>. When preincubated, preincubation time was 7 minutes. Curve 1. 1 mM ATP added to preincubation mixture, no further addition. Curve 2. Preincubation with $2 \cdot 10^{-5}$ M ATP, 1 mM ATP added at the start of the incubation. Curve 3. Preincubation with 10^{-5} M ATP, 1 mM ATP added at the start of the incubation. Curve 4. No preincubation. 1 mM ATP added. Curve 5. Preincubation with $2 \cdot 10^{-5}$ M ATP. 1 mM ITP added at the start of the incubation.
10^{-5} M ATP or $2 \cdot 10^{-5}$ M ATP respectively added to the preincubation mixture was with no further addition unable to give any reduction. Further experimental conditions as in Fig. 1.

Under the conditions used for the reduction of DPN, the splitting of high-energy bonds, measured as the formation of inorganic phosphate, is also greatly influenced by preincubation. In Fig. 5 the rate of P_i-release is plotted against time of preincubation. As will be seen, preincubation had no effect when there were no additions to the basic system or in the presence of succinate alone. On the other hand, 3 minutes of preincubation gave an increase of 75 % in the rate of P_i-release when both succinate and DPN were added. DPN alone caused no stimulation.

The same picture can be obtained when TMPD-ascorbate is the source of electrons.

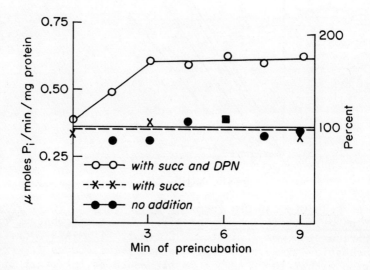

Fig. 5 <u>Effect of preincubation on the ATPase activity.</u>
Preincubation was attained, for times indicated, in a medium with Tris buffer; $MgCl_2$; sucrose; ATP and amount of protein as in Fig. 1. The incubation was started by addition of succinate or succinate and DPN and the release of inorganic phosphate compared to that where no addition was made was determined after 5 minutes. Release of P_i during preincubation subtracted. Determination of P_i according to Ernster <u>et al</u>. (1950). Final volume was 3.0 ml and the reaction was followed at $30°$ C.

If the energy requirement for the reduction is calculated on the basis of the total release of inorganic phosphate per DPNH formed, one finds a requirement of 3-4 high energy bonds per DPN reduced from succinate, and between 5-6 when DPN is reduced from the cytochrome c level, as seen in Table 2.

TABLE 2

Determination of the P_i/DPNH quotient

Experimental conditions as in Fig. 1 except that the reference cuvette contained the complete reaction mixture without DPN. The particles were preincubated for 7 minutes. After an incubation time of 4 minutes during which the absorption at 340 mµ was followed, aliquots were simultaneously taken out from each of the two cuvettes and analysed for P_i liberated. Column 1 gives the amounts of DPN reduced, columns 2 and 3, the amount of P_i liberated with and without DPN added. The next column gives the P_i/DPNH quotient based on the total release of P_i and the last column the quotient between the increase in P_i release caused by the addition of DPN to the amount of DPN reduced.

Electron source	1 µmoles DPNH	2 µmoles P_i compl.	3 µmoles P_i no DPN	$\frac{2}{1}$	$\frac{2-3}{1}$
Succinate	0.113 0.116	0.407 0.371	0.272	3.6 3.1	1.2 0.9
TMPD and Ascorbate	0.069 0.060	0.355 0.354	0.237	5.2 5.9	1.7 2.0

This energy requirement can also be calculated on the basis of the stimulation of the P_i-release caused by the addition of DPN (cf. Fig. 4). Under such circumstances one obtains a value of 1 high energy bond split for the reduction of one molecule of DPN when succinate was used as the reducing agent. The corresponding figure when the reduction was carried out by ascorbate and TMPD was 2. These values are indeed of great interest, but before they are accepted as "backwards" P/O ratios the specificity of the DPN stimulation of the P_i-release must be better established.

As I hinted earlier in my talk, ITP can be used as a source of energy. Fig. 6 shows that 1 mM ITP gives a very low rate of reduction of DPN by succinate, but this rate is

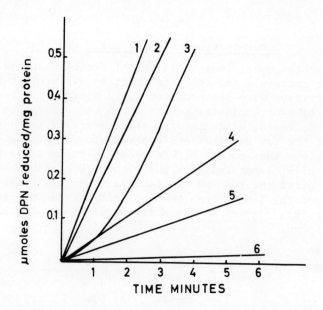

Fig. 6. <u>The effect of ITP on the succinate linked DPN reduction</u>. Preincubation time when indicated 7 minutes. Curve 1. 1 mM ATP added to the preincubation mixture, no further addition. Curve 2. Preincubation with 1 mM ITP, 1 mM ATP added at start of the incubation. Curve 3. No preincubation. 1 mM ATP added. Curve 4. 1 mM ITP added to the preincubation mixture, a further addition of ITP to 2 mM final concentration at the start of the incubation. Curve 5. Preincubation with 1 mM ITP, no further addition. Curve 6. No preincubation. 1 mM ITP added.

increased by preincubation and if a further addition of ITP is made after the preincubation the rate of reduction becomes considerable. Fig. 6 also demonstrates clearly that ITP also abolishes the lagtime when the reduction is driven by a second addition of ATP.

The picture is the same when TMPD and ascorbate is used as the electron source (Fig. 7). ITP without preincubation gives little or no reduction. Preincubation with ITP not only abolishes the lag period but also leads to an increase in rate, when the reduction is driven with ATP. The dashed lines in the figure represent the best rate of reduction with

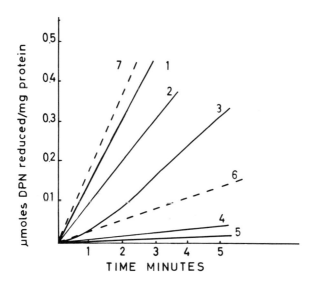

Fig. 7. <u>The effect of ITP on the reduction of DPN from the cytochrome c level.</u> Preincubation time when not otherwise indicated 7 minutes. Curve 1. 1 mM ATP added to the preincubation mixture, no further addition. Curve 2. Preincubation with 1 mM ITP. 1 mM ATP added at the start of the incubation. Curve 3. No preincubation. 1 mM ATP added. Curve 4. Preincubation with 1 mM ITP. a further addition of ITP to a final concentration of 2 mM at start of the incubation. Curve 5. 1 mM ITP added to the preincubation mixture, no further addition. Curve 6. Preincubation with 1 mM ITP, at the start of the incubation a further addition of ITP to 2 mM final concentration. The amount of Mg^{++} raised to 8 mM from the usual 6 mM. Preincubation time 8 minutes. Curve 7. Preincubation with 1 mM ATP, no further addition. Final concentration of Mg^{++} 8 mM.

ITP which we have obtained with this system so far. It was obtained with a somewhat prolonged preincubation and with 8 mM Mg^{++} present instead of 6 mM as was usually used in these experiments.

ITP has been used as an example of a nucleoside triphosphate other than ATP. GTP and UTP work almost as well, but CTP is without effect. This is in accordance with what has been reported for the isolated ATPase by Dr. Racker's group (4).

TABLE 3

Effect of various phosphate acceptors on oxidative phosphorylation in submitochondrial particles from beef heart

The incubation medium contained 50 µM glycylglycine buffer pH 7.5; 10 µM succinate; 10 mM $MgCl_2$; 12.5 mM phosphate buffer pH 7.5; 0.25 M sucrose. The various nucleoside diphosphates were added to a concentration of 5 or 10 mM. Glucose to 30 mM and hexokinase corresponding to 0.025 mg of crystalline material per ml when indicated. Particles added equal to 1 mg of protein/ml. The final volume was 1.0 ml and the reaction was followed for 20 minutes at 30° C.

Phosphate acceptor	µatoms O	µmoles P	glucose + hexokinase	
			µatoms O	µmoles P
----	5.9	0.0	-	-
ADP 5 µmoles	5.9	3.7	-	-
CDP 5 µmoles	5.3	0.2	-	-
GDP 5 µmoles	7.3	3.2	-	-
IDP t µmoles	7.1	3.4	-	-
UDP 5 µmoles	6.5	2.0	-	-
ADP 10 µmoles	6.4	3.8	6.2	4.9
GDP 10 µmoles	6.9	2.8	5.4	2.1
IDP 10 µmoles	5.7	2.8	6.0	3.7

Since oxidative phosphorylation has always been regarded as a specific phosphorylation of ADP it appeared that this was a reversal which had nothing to do with oxidative phosphorylation. To our surprise, when we tested for it (Table 3) we found that the oxidative phosphorylation in these particles was by no means specific for ADP as acceptor but also worked remarkably well with all tested nucleoside diphosphates except CDP. On the other hand in intact mitochondria the reaction proved to be specific for ADP (Table 4). The phosphorylation of IDP in particles was inhibited by dinitrophenol and by oligomycin and it disappeared when the respiration was inhibited by antimycin A (Table 5).

TABLE 4

Effect of various phosphate acceptors on oxidative phosphorylation in beef heart mitochondria

The incubation medium contained 2 mg of mitochondrial protein/ml, 10 mM $MgCl_2$; 25 mM phosphate buffer pH 7.5; 0.25 M sucrose and 20 mM succinate giving a final volume of 1.0 ml. Incubation time 20 minutes at 30° C.

Phosphate acceptor	μatoms O	μmoles P	glucose + hexokinase	
			μatoms O	μmoles P
ADP 5 μmoles	2.4	4.9	2.4	4.9
----	3.5	0.5	5.1	1.1
CDP 5 μmoles	3.3	0.2	3.3	0.9
GDP 5 μmoles	4.6	0.3	4.4	0.9
IDP 5 μmoles	4.9	0.6	4.6	1.1
UDP 5 μmoles	3.2	0.4	3.9	1.1

TABLE 5

Effect of various inhibitors on oxidative phosphorylation with IDP as phosphate acceptor in submitochondrial particles

Incubation medium similar to that in Table 3 without addition of hexokinase-glucose. Oligomycin and antimycin A were added as 10 μl aliquots of alcoholic solutions. Final volume 1.0 ml. Incubation time 20 minutes at 30° C.

Addition	μatoms O	μmoles P
----	5.7	3.1
DNP, 10^{-4} M	5.9	0.3
antimycin A, 1 μg/mg prot.	0	0.1
oligomycin, 0.5 μg/mg prot.	5.7	0.1

From the present data, however, it seems obvious that the specific ATP reaction comes in at a late stage and might be formulated

$$NTP + ADP \rightleftharpoons NDP + ATP$$

where N stands for A, I, U or G. This type of kinase reaction has long ago been suggested by Siekevitz and Potter (5) to be the transport mechanism of high energy bonds in intact mitochondria.

The "classical" theory of an activation of P_i, rather than activation of the nucleotide moiety, rests on the finding that in ATP formed by intact mitochondria, the terminal bridge oxygen is derived from the ADP (6), not from the inorganic phosphate.

When the above formulated reaction is introduced as the terminal transfer reaction of high energy bonds in intact mitochondria, there might be reasons to reopen the discussion of whether the nucleotide moiety, NDP, or the P_i is the part that is activated.

It should be possible to get a direct answer by studying the $ADP-H_2O^{18}$ - exchange in our system. The discussion of these findings in the light of earlier concepts developed on the basis of various exchange reactions must be postponed until more experiments have been carried out.

REFERENCES

1. Löw, H., Krueger, H. and Ziegler, D.M., Biochem. Biophys. Res. Com., 5, 231 (1961).
2. Löw, H. and Vallin, I., Biochim. Biophys. Acta, 69, 361 (1963).
3. Löw, H. and Vallin, I., Biochem. Biophys. Res. Com., 9, 304 (1962).
4. Pullman, M.E., Penefsky, H.S., Datta, A. and Racker, E., J. Biol. Chem., 235, 3322 (1960).
5. Siekevitz, P. and Potter, V.R., J. Biol. Chem., 215, 237 (1955).
6. Boyer, P.D., Proc. Intern. Symp. Enzyme Chem., Tokyo and Kyoto, Maruzen and Co. Ltd., London, 1957, p. 301.
7. Ernster, L., Zetterström, R. and Lindberg, O., Acta Chem. Scand., 4, 942 (1950).

DISCUSSION

Hess: Are these particles all sonicated?

Löw: Yes, we have been using particles derived from beef heart mitochondria by sonic treatment.

Chance: It seems to me that much that occurs in sonicated particles is the opposite of what happens in intact mitochondria, and that you have underlined this beautifully. One point is that ITP is the only inhibitor of the ATP-activated DPN reduction in mitochondria; the other point is that DPN reduction activates ATP-ase in your case, while it inhibits in the mitochondria. Thus there are many contrasts. Instead of a diphospho-kinase in the mitochondria, I would suggest that ATP-ase itself has somehow been altered in its nucleotide specificity. It seems to me that there is nothing you have said which would distinguish between these two hypotheses.

Löw: No, there is not, and although that is a possible explanation, I would like to point out that atractylate, which is known to act like oligomycin in intact mitochondria and also in digitonin particles (1,2) is not, as you can see from Table I, acting at all with any substrate in sonicated particles. So the atractylate-sensitive reaction in sonicated particles seems to be lost, which may indicate that there is one reaction step lost.

Racker: Do you require the same concentrations of each nucleotide in oxidative phosphorylation? Do you need as little IDP as ADP? Or do you need much more?

Löw: We used as little IDP as ADP.

Hommes: Perhaps the problem of the ATP-dependent cytochrome b reduction, which is inhibited by malonate when succinate is present as electron donor, has been resolved in the scheme where Dr. Sanadi shows a different pathway of electron transfer. (see p. 29).

Löw: Yes, that might be possible. We were very much puzzled by your results.

TABLE I

EFFECT OF ATRACTYLOSIDE AND OLIGOMYCIN ON
OXIDATIVE PHOSPHORYLATION IN SUBMITOCHONDRIAL PARTICLES

The incubation medium contained 50 mM glycylglycine pH 7.5;
2 mM $MgCl_2$; 0.25 M sucrose; 12.5 mM phosphate buffer pH 7.5;
1 mM ATP; 30 mM glucose and hexokinase corresponding to 0.025
mg of crystalline material. Particles added equal to 1 mg of
protein/ml. 10 mM succinate or $3.3 \cdot 10^{-4}$ M DPN; 10^{-3} M semi-
carbazide; $5 \cdot 10^{-4}$ mg of ADH/ml and a final volume of 10 μl
ethanol including that with the dissolved oligomycin. The
reaction was followed for 20 minutes at 30° C.

addition	succinate		DPN + ADH	
	uatoms O	umoles P	uatoms O	umoles
---	6.6	5.6	3.0	5.0
atractyloside 5 x 10^{-5}M	5.7	4.6	3.0	5.7
-"- 10^{-3}M	5.2	5.0	2.0	5.1
oligomycin 0.5 μg/mg prot.	5.6	0.2	2.2	0.3
" 1.0 μg/mg prot.	4.9	0.4	2.1	0.2

<u>Klingenberg</u>: I wanted to come back to the difference in the
effects of ITP and ATP on the phosphorylation potential for
different preparations. The ATP effect is inhibited by magnesium in mitochondria, while in the particles it is activated
by magnesium. Does ITP have a different binding constant for
magnesium than ATP? This may be related to the different magnesium requirements for ITP and ATP effects on your particles.

<u>Löw</u>: As you saw, the dashed line in my Fig. 7 was our best
experiment with ITP. There we used 8 mM magnesium for ITP,
instead of the 6 mM magnesium that was maximal for ATP. So
the difference is not very great. We expected that ITP would
have a higher magnesium requirement. In this connection, I
should point out that preincubation does not require magnesium
of the same order of magnitude as does reduction, but rather
something like 10^{-5} M.

<u>Chance</u>: Isn't it true that this figure, 10^{-5}M, may well be of
the order of the magnesium binding constant for ADP? I don't
believe in any case that there would be enough difference to
explain the difference of activity. The difference of specificity might well raise questions as to what has happened to

the histidine phosphate, which is known to be formed in mitochondrial systems--but I have heard from Madison that it is not found in ETPH.

Webster: Initial experiments were negative, but Dr. Boyer tells us he has now found small amounts of histidine phosphate formed by ETPH.

Chance: Referring to the Antimycin sensitivity of reversed electron transfer, the experiments that Hollunger and I carried out initially were more of a survey type of experiment since, at that time, we did not anticipate that the exact concentration of Antimycin during inhibition of reversed electron transfer would arouse as much interest as indeed, it has. In that paper we simply obtained striking effects upon the steady state of reduced pyridino nucleotide with concentrations of 3γ per milligram protein. Such concentrations immediately reversed the effect of succinate and ATP on DPN reduction. This value, of course, was a maximum level (3). A more recent study of the effect of Antimycin upon ATP-ase of pigeon heart mitochondria has appeared (4). For this reason, in order to obtain a more quantitative result, Dr. C. P. Lee and I have very recently re-examined this question.

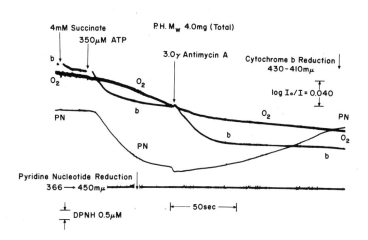

Fig. 1 (Chance and Lee). Our experimental procedure is indicated in Fig. 1. Pigeon heart mitochondria suspended in mannitol-sucrose-tris medium are supplemented with 4 mM succinate. Succinate alone causes negligible cytochrome b reduction (top trace, lower case b), oxygen utilization (middle

trace, label O_2), or pyridine nucleotide reduction (lowest trace, label PN). Addition of 350 µmolar ATP causes immediate cytochrome b reduction, a slightly delayed pyridine nucleotide reduction, and a considerably delayed activation of respiration. The mitochondria are now in State 4. Addition of 3γ of Antimycin to a cuvette containing 4 mg protein (0.75γ Antimycin/mg protein) causes an immediate increase in the reduction of cytochrome b, approximately 50 per cent inhibition of respiration, and a reversal of DPN reduction (the fluorescence trace rises slowly so that the final level of reduction is almost negligible).

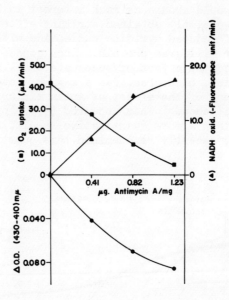

Fig. 2 (Chance and Lee). A series of these experiments has been carried out and the results are plotted in Fig. 2. Half-maximal inhibition of oxygen utilization is obtained with 0.42 γ Antimycin/mg protein; half-maximal cytochrome b oxidation is obtained with 0.31γ Antimycin/mg protein. In the case of pyridine nucleotide, the most conservative criterion for the inhibitory effect of Antimycin was taken in terms of the acceleration of the disappearance of DPNH as a function of the concentration of Antimycin (a very small amount of Antimycin will cause the fluorescence to disappear in several minutes, as indicated by Fig. 1). The figure shows that 0.4γ Antimycin

A/mg protein is half-maximally effective in this reaction. On the basis of these data we are able to draw the conclusion that the concentrations of Antimycin A affecting cytochrome b and State 4 respiration in pigeon heart mitochondria are very closely related or about equal to those causing half-maximal effects of energy-linked pyridine nucleotide reduction. Thus the value previously published (less than 3γ/mg protein) is surely a maximal value and can be revised downwards toward 0.5γ/mg protein. (Note added in proof: a point which the author failed to consider in the discussion is the metabolic state of the mitochondria. Our data on respiratory inhibition, cytochrome b oxidation, and pyridine nucleotide reduction are based upon State 4, where mitochondria are much less sensitive to inhibitors in general than in State 3. The data which Professor Slater reports for 0.1γ/mg protein are presumably for State 3).

Slater: Dr. Chance, may I comment on your study of the effects of Antimycin? In our experiments with liver mitochondria, the ATP-dependent reduction of DPN by succinate (measured by glutamate formation) is considerably inhibited by 0.75γ Antimycin/mg protein, exactly the same concentration which you used. But about 0.1γ/mg protein is sufficient for complete inhibition of forward electron transfer in the respiratory chain in mitochondria. There is really a rather critical area between 0.1 and 0.75γ Antimycin/mg protein. I would agree completely with Löw that concentrations of Antimycin which are sufficient to inhibit the respiratory chain do not inhibit the reversed electron transfer. You do start getting inhibition at higher concentrations, as you have shown. This may be because Antimycin is also an uncoupler.

Chance: Our systems may differ. The respiration which we have examined with pigeon heart mitochondria is rather labile; the electron donor here was succinate in the presence of ATP which, as a matter of fact, is not the most inhibitor-resistant system--it is observed to be inhibited even by ADP. What you are saying is that the effect on respiration should be measured in the presence of succinate and glutamate, conditions similar to your own. Have I understood your question correctly?

Slater: No. My question is: is the ordinary Antimycin-inhibited site involved in reversed electron transfer to DPN from succinate? Dr. Löw's data would say no, because he finds no inhibition of reversed electron transfer by concentrations of Antimycin which do inhibit the respiratory chain. You get some inhibition at a bit higher concentrations; Löw gets inhibitions at much higher concentrations. I would agree with Löw completely on this point. The ordinary Antimycin-sensitive site is not involved in the succinate-DPN reaction.

Chance: I would subscribe to the view that this is true in sonicated particles.

Slater: But we get the same results as Löw using intact liver mitochondria.

Klingenberg: The question is whether the primary action of Antimycin might be on electron transfer or energy transfer. In the whole system, only a secondary effect of Antimycin may affect the energy transfer, by inhibiting in a primary effect the electron transfer in the cytochrome system. For example, added ATP can thus no longer supply energy via the energy transfer pathway of the cytochrome region, in order to preserve an organization which is necessary for the succinate-DPN reduction. It might be important to the mitochondria to keep up this organization where DPN must be located in a certain way to be rapidly reduced by succinate.

Estabrook: I have a question about the validity of your calculations of ATP:DPNH ratios. This concerns the question of the subtraction of the ATP-ase. If you were to calculate P:O ratios as determined by oxygen stimulation on ADP addition along the same lines, you could get a P:O ratio of 10. Are you concerned about this correction? It is hard to determine its validity.

Löw: Upon addition of DPN there is a certain stimulation of the ATP-ase. If one makes the assumption that this stimulation of ATP-ase has to do with the reduction of the added DPN, one could calculate P_i:DPNH ratios on this basis. Whether or not this assumption is valid is difficult to say as yet. It would seem that the ratios calculated on this basis agree too well with theory to be true.

Mildvan: Is there an ITP-ase or a GTP-ase activity which is less active, and would therefore afford a simpler evaluation of the P_i:DPNH ratio?

Löw: This is indeed a very good possibility. We have started experiments of this type.

Chance: Dr. Racker, what are the relative activities of F_1 towards the two nucleotides?

Racker: ITP is actually hydrolyzed about 50 per cent faster than ATP by the soluble mitochondrial ATP-ase (F_1).

ENERGY-LINKED FUNCTIONS OF MITOCHONDRIA

Low: In our particles, ITP is hydrolyzed at about 50 per cent of the rate for ATP. One peculiarity is that so far we have had difficulties in demonstrating any ITP-P_i exchange; there is some, but the ATP-P_i exchange is about 10 times faster.

Racker: May I point to one difference in the reactivity of ATP-ase to the nucleotides which may be critical here? The enzyme is susceptible to inhibition by ADP but not by the other nucleotide diphosphates. This may explain the greater efficiency of the adenine nucleotides in the forward reaction.

Klingenberg: What is the ratio between the DPN that cannot be reduced by succinate and protein content? Is this linear?

Hommes: Yes, there is a linear relationship, which depends a little on the particular preparation. In general, it is in the range of 0.5 to 1.0 mµM DPN/mg protein.

Estabrook: This is roughly one per cytochrome.

Klingenberg: Thus the amount one can reduce with succinate is about the same as one would have in mitochondria?

Hommes: At this low DPN level, yes; at higher DPN concentrations the ATP-ase activity of the particles interferes with this type of experiment.

Slater: May I ask a related question of Dr. Sanadi? In your system, following NAD reduction you tested with alcohol dehydrogenase and acetaldehyde to find if it was all NADH. Does the NADH absorption come down to zero?

Sanadi: No, it does not. I could not account for 5-10 per cent of the absorption change. Apparently, there is a nonenzymatic reaction between DPN and reduced quinone with a maximum at 330 mu that goes on at the same time. Those blanks we saw without ATP and without enzyme refer to this. This change generally amounts to anywhere between 0.02 and 0.04 in five minutes.

Packer: We have been speaking a good deal about the differences between submitochondrial particles and mitochondria, and it may be that the mitochondrial system is much simpler, and therefore the results obtained are more valid with respect to the basic mechanism. However, with ascorbate and tetramethylphenylene diamine we have never been able to demonstrate in the intact mitochondria an ATP dependence for pyridine nucleotide reduction under conditions where it is very readily demonstrated with succinate. Would you comment on this?

Löw: Is that in a terminally inhibited system?

Packer: No, there is no cyanide present.

Löw: Yes, but as far as I know, you have demonstrated oxidative phosphorylation from TMPD to oxygen. Couldn't this handle the supply of high energy?

Packer: We do this in a phosphate-depleted system.

Löw: It is not necessary to have phosphate. As has been shown by Ernster and his co-workers (5), the reaction may follow an oligomycin-insensitive path; it is possible to drive electrons from succinate to DPNH by using the oxidation of succinate by oxygen in the presence of oligomycin as a source of energy. However, there is a certain difference between mitochondria and submitochondrial particles, because from your experiment with mitochondria one would expect that TMPD should be a beautiful substrate for oxidative phosphorylation in the cytochrome oxidase region, but in our hands, using sonicated particles, it is not. One can demonstrate an oxidative phosphorylation with reduced TMPD but, for instance, reduced phenazine methosulfate is a much better substrate to demonstrate oxidative phosphorylation, whereas TMPD is a better substrate for the energy-requiring reduction of DPN.

Racker: What seems to bother Dr. Packer is that in one case ATP works and in the other it does not. This could be a difference in the accessibility of the different sites to ATP, which may also explain the lack of ATP-ase activity in the system described by Green and his collaborators (6).

Packer: I find that hard to believe.

Pullman: It is not that the system cannot use ATP, but merely that you cannot show a requirement for it in an aerobic system.

Packer: But you can show the requirement for the succinate system in the same aerobic experiments.

Slater: He is making $\sim$ compounds from oxidative phosphorylation, which is easier than making them from ATP.

Packer: Why can't you make them from succinate, too?

Slater: You can make them from succinate.

Packer: But you do not get the reduction with succinate unless you add ATP, whereas you do not have to add ATP for the TMPD system.

Conover: Is the reaction carried out in State 3 or State 4?

Packer: It is a phosphate-free system.

Chance: It has, in any case, something to do with energy conservation at the succinate site. So perhaps we do have an intact site from cytochrome c to oxygen.

Packer: Perhaps we can discuss this further after my paper.

Estabrook: I would like to bring up another point which may be clarified by the end of the afternoon; the absolute magnesium requirement for submitochondrial systems as compared with the inhibitory effects of magnesium on mitochondrial systems. Dr. Chance, do you believe that the intermediate is used to transport magnesium into the mitochondria, instead of expending the energy in the reduction of pyridine nucleotide--or how do you explain this?

Chance: We attribute the inhibition of reversed electron transfer by added magnesium to a binding of ATP with magnesium. We suggest that the transfer enzyme of the mitochondria is sensitive to the binding state of the ATP, while the enzyme of the sonicated particles is insensitive to the binding state, or even the nucleotide moiety. Although this may be an oversimplification, it is consistent with your data on ITP. I imagine that the native transfer enzyme reacts only with free ATP; that after sonication the transfer enzyme reacts with magnesium-bound ATP or with bound ITP. This also appears to be true of F_1, the cold-labile ATP-ase which has "lost" its specificity for ATP. We do not know whether these different specificities preexist, or whether there is a change in specificity that occurs on sonication; perhaps a new site could be opened up. One would expect that the charge properties of magnesium-ATP might be inappropriate to the enzyme in one state and not be a matter of relevance to the enzyme in the other state. In summary, I do not think that one must postulate an additional kinase at this point, but, of course, it is certainly not ruled out.

REFERENCES

1. Bruni, A., Contessa, A. R., and Lucianin, S., Biochim. Biophys. Acta, $\underline{60}$, 301 (1962).

2. Vignais, P. V., Vignais, P. M., and Stanislas, E., Biochim. Biophys. Acta, $\underline{60}$, 284 (1962).

3. Chance, B., and Hollunger, G., J. Biol. Chem., $\underline{236}$, 1564 (1961).

4. Chance, B., and Ito, T., J. Biol. Chem., $\underline{238}$, 1509 (1963).

5. Ernster, L., in Symposium on Intracellular Respiration, Vth Internatl. Congr. Biochem., Moscow, 1961. Pergamon, Oxford, 1963, $\underline{V}$, 115.

REDUCTIVE DEPHOSPHORYLATION OF ATP IN THE DPNH-COENZYME Q SEGMENT OF THE RESPIRATORY CHAIN

D. R. Sanadi, Thomas E. Andreoli, Richard L. Pharo, and S. R. Vyas

Gerontology Branch, National Heart Institute, National Institutes of Health, PHS, U. S. Department of Health, Education & Welfare, Bethesda, and the Baltimore City Hospitals, Baltimore, Maryland

The two reactions that I wish to discuss today may be represented by the Equations 1 and 2.

$$\text{Ascorbate} + DPN^+ + ATP \xrightarrow{\text{Quinone}} \text{Dehydroascorbate} \atop DPNH + H^+ + ADP + P_i \quad (1)$$

$$DPNH + H^+ + \text{Quinone} \longrightarrow DPN^+ + \text{Hydroquinone} \quad (2)$$

The first reaction is the ATP-requiring reduction of DPN by ascorbate in the presence of catalytic amounts of a quinone which may be menadione (vitamin K_3) or coenzyme Q_1 (CoQ_1). The second reaction is the oxidation of DPNH by menadione, CoQ_1 or CoQ_6 catalyzed by a highly purified enzyme. It is our hypothesis that the second reaction may be the electron transfer portion of Equation 1.

These reactions are catalyzed by cyanide inhibited submitochondrial particles (ETP_H) from beef heart. The course of Reaction 1 may be followed by measuring the increase in absorption at 340 mµ due to DPNH formation. As shown in a preliminary communication, the initial rates of DPN reduction observed with menadione or CoQ_1 as the catalytic electron carrier are identical to the rates obtained with succinate as the reductant (1). The absorbancy change has been established to be due to the production of DPNH by using alcohol dehydrogenase and acetaldehyde which reverse the change almost completely. In the absence of ATP, quinone, DPN or enzyme, the rate of increase in absorbancy is extremely slow (Table I). A lag period, varying with the assay conditions and particle preparation, has been observed between the addition of the final reactant and the onset of DPN reduction.

TABLE I

ATP-dependent reduction of DPN by ascorbate

System	$\Delta A_{340\ m\mu}/8$ min.	
	Coenzyme Q_1	Menadione
Complete	0.422	0.411
No ATP	0.064	0.046
No quinone	0.030	0.019
No ascorbate	0.027	0.018
No DPN	0.030	0.021
No enzyme	0.043	0.056

The ATP-dependent reduction of DPN by ascorbate with ETP_H is inhibited strongly by reagents that uncouple or inhibit oxidative phosphorylation. Nearly complete inhibition is obtained with 0.2 mM 2,4-dinitrophenol, 0.1 mM dicumarol and 0.5 μg oligomycin/ml of reaction medium. Low levels of Amytal (1.5 mM) and trifluorothienylbutanedione (0.1 mM), which inhibit respiration by affecting the DPNH dehydrogenase site also inhibit DPN reduction. The results

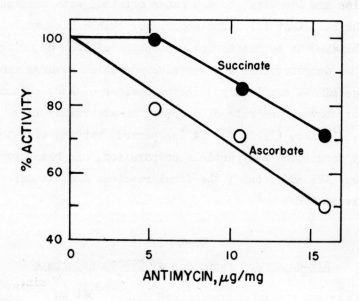

Fig. 1. _Inhibition of energy-requiring DPN reduction in the presence of antimycin A._

with antimycin A are shown in Fig. 1; 50 per cent inhibition is obtained at inordinately high levels just as in the succinate-linked DPN reduction system. It would appear that, at least in these particles, the classical antimycin site operating between cytochromes b and c_1, is not involved in DPN reduction by succinate or ascorbate and menadione. On the basis of these results, the pathway of electron flow in Reaction 1 may be represented as in Fig. 2. The hydroquinone produced by non-enzymatic reduction by ascorbate may reduce either cytochrome b or CoQ or the DPNH dehydrogenase flavoprotein. The hydrogens may then be transferred to DPN

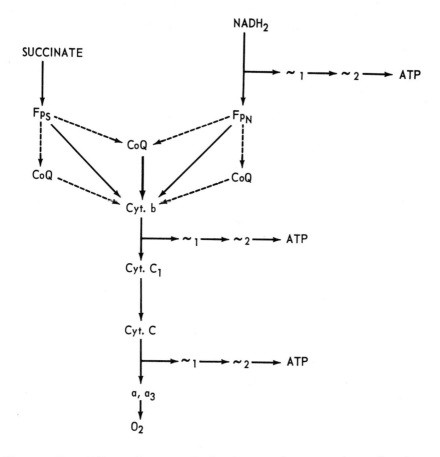

Fig. 2. <u>Possible pathways of electron and energy transfer in the energy requiring DPN reduction</u>

at the expense of the energy supplied by ATP via the energy coupling reactions associated with the oxidation at the DPNH site.

We had reported earlier (2) that the rate of reduction of DPN by succinate in mitochondrial particles is reduced greatly if the particles are washed. The activity is restored by the addition of a 100,000 x g supernatant fraction from sonicated mitochondria. The fraction also stimulates the esterification of phosphate coupled to the oxidation of DPNH by fumarate in cyanide inhibited particles (3). A similar situation exists in the ascorbate-mediated reduction of DPN. The activity of the washed particles is stimulated over three-fold by a soluble supernatant which by itself is completely inactive (1).

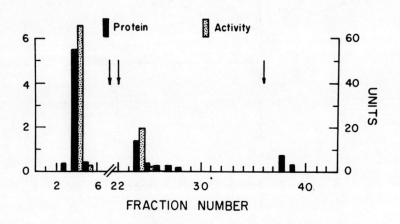

Fig. 3. Resolution of DPNH-coupling factor on DEAE-cellulose

A resolution of the coupling activity occurs when the coupling factor, purified approximately 3 to 6-fold by fractionation with ammonium sulfate, is chromatographed on a column of DEAE-cellulose. Under these conditions the activity comes off in two distinct peaks as seen in Fig. 3. The first peak is only slightly retarded in 5 mM Tris sulfate while the second peak comes off at approximately 0.25 M Tris sulfate. Fig. 4 shows the effect of combining the proteins from the two peaks. The broken line represents the arithmetic sum of the individual activities of Peaks I and II.

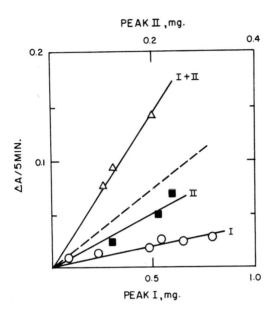

Fig. 4. Effect of combination of active fractions

When the two fractions are combined in the assay, the observed activity, represented by the triangles, is at least twice as high as the sum of the individual activities. This type of data has been obtained repeatedly and we are inclined to interpret it to mean that the two fractions are functionally distinct in the ATP-dependent reduction of DPN by succinate.

Table II shows that both coupling fractions have significant ATPase and pyrophosphatase activities (4). The relationship between these activities and the coupling function remains to be clarified. I might point out that although the apparent purification of Peaks I and II appears

TABLE II
ATPase and pyrophosphatase activity of coupling factors

	μmoles/min/mg		
	DPNH	ATPase	PPase
Peak I	0.20	0.22	0.12
Peak II	0.18	0.33	0.28

to be only 5-fold, the actual purification is probably considerably higher since the crude extract would show the individual activities of the two peaks as well as the added stimulation when both are present as indicated in Table II.

These mitochondrial particles are also capable of catalyzing the oxidation of DPNH by menadione and by CoQ_1 as reported (5) and in addition by CoQ_6. The activities are inhibited by 0.5 mM Amytal and 0.1 mM trifluorothienylbutanedione. Antimycin A is also an inhibitor but the concentration necessary for 50 per cent inhibition is as high as in the reduction of DPN by ascorbate (20 μg/mg protein). These results support the suggestion that antimycin is capable of acting at a second site different from the more sensitive site between cytochromes b and c_1.

TABLE III
Activities of particulate and purified DPNH-CoQ_6 reductase

	Electron acceptor	Total units	Units/mg	Purification
Particles	Menadione	394	0.48	
	CoQ_6	502	0.65	
Purified	Menadione	660	98.5	205 fold
	CoQ_6	107	17.5	27 fold

TABLE IV

Properties of purified DPNH-CoQ reductase

Sedimentation value	5.7 s
Flavin component	
FMN - μ moles/gm (approx.)	10
Min. mol. wt. (approx.)	100,000

A soluble flavoprotein with high DPNH-menadione and CoQ_6 reductase activities has been extracted from the particles and purified on hydroxyapatite (Table III). The unit here represents 1 μ mole of DPNH oxidized per min. at 30°. The activities of 1 μg and 3 μg purified enzyme in menadione and CoQ_6 reduction respectively in 1 ml reaction medium are illustrated in Fig. 5. The reaction goes to completion when

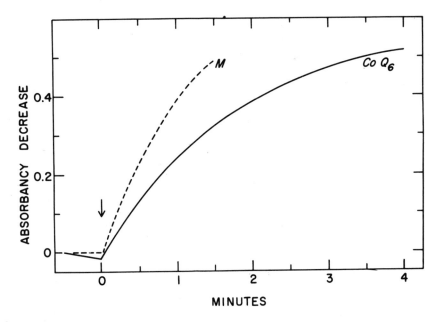

Fig. 5. DPNH oxidation by menadione and CoQ_6 catalyzed by purified reductase

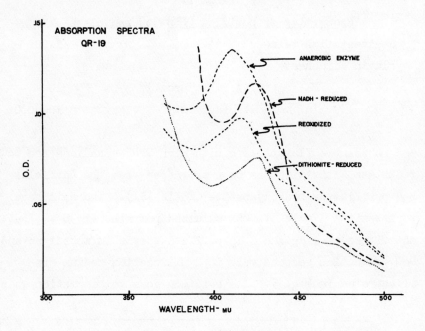

Fig. 6. Absorption spectrum of DPNH-CoQ$_6$ reductase under different conditions

any one of the components is limiting in the medium. Some of the properties of the enzyme are shown in Table IV. The flavin prosthetic group, which is readily extracted by boiling and acidification, is indistinguishable from FMN by paper chromatography in two solvent systems. The minimum molecular weight of the enzyme, based on the flavin content is approximately 100,000. The low sedimentation constant (5.7 s) would also indicate that the flavoprotein is of low molecular weight. Its absorption spectrum shows a prominent absorption peak at 408 mμ and a pronounced shoulder at 450 mμ (Figure 6). On addition of DPNH, there is considerable bleaching at both wave lengths. The difference spectrum, oxidized minus reduced, resembles the spectrum of flavin in the 450 mμ. The enzyme is bleached further, particularly at 408 mμ, by dithionite.

The investigations on the DPNH-CoQ$_6$ reductase are still at an early stage and the results must be considered as preliminary reports.

REFERENCES

1. Sanadi, D. R., J. Biol. Chem., 238, PC 482 (1963).
2. Sanadi, D. R., Fluharty, A. L., and Andreoli, T. S., Biochem. Biophys. Res. Comm., 8, 200 (1962).
3. Sanadi, D. R., and Fluharty, A. L., Biochemistry, in press.
4. Racker, E., Proc. Natl. Acad. Sci., 48, 1659 (1962).
5. Hatefi, Y., Haavik, A. G., and Griffiths, D. E., J. Biol. Chem., 237, 1676 (1962).

DISCUSSION

<u>Pullman</u>: Are either of these two coupling factors cold-labile?

<u>Sanadi</u>: Neither is cold-labile; incubation at $30°$ for 30 minutes will destroy both activities about 80-90 %, and at $0°$ for one hour there is no detectable loss of either activity.

<u>Pullman</u>: Did you follow the ATPase activity?

<u>Sanadi</u>: No, we have followed only the coupling activities.

<u>Racker</u>: In spite of the fact that you said it was not cold-labile, I recommend that you try ATPase. Dr. Prairie in our laboratory tried the repeated washings which you described, and he finds that he can restore some of the activity by adding purified ATPase.

<u>Sanadi</u>: Our purification has not advanced to the point where we can be sure of ATPase involvement or not, in our particular system with two components.

<u>King</u>: Does this contain Fe?

<u>Sanadi</u>: We have not done the analysis yet. I would not be surprised because of the 408 mμ absorption peak.

<u>King</u>: Are you sure it is FMN and not FAD?

<u>Sanadi</u>: Yes, of that I am convinced. The fluorescence yield also corresponds to FMN, although we have not done this by refined methods.

<u>King</u>: Very good!

<u>Estabrook</u>: Would you comment on Löw's statement that his quinone reductase system was Antimycin-sensitive, whereas yours was Antimycin-insensitive on a relative scale?

<u>Sanadi</u>: I arrived late; what was his observation, please?

<u>Löw</u>: We have not found any Antimycin-insensitive mediators of electrons into the chain, other than succinate. We have tried methyl-Q_0.

<u>Sanadi</u>: You find that Q_0 works?

<u>Löw</u>: Yes, if instead of a "zero" you put a methyl group, it is sensitive to Antimycin.

Sanadi: That may be a reaction similar to menadione and Q_1 in our system.

Low: Is your Q_1 Antimycin-insensitive?

Sanadi: Yes, it is insensitive except at very high levels. The purified Q-reductase is also sensitive to Antimycin, but at inordinately high levels: about 1 mg Antimycin/mg protein gives about 30 % inhibition.

Slater: Returning to Estabrook's last point: is there a discrepancy with Q_1?

Low: I have used Q_1, but for Antimycin sensitivity we have tested the methyl Q_0 only.

Slater: But you do get very good Antimycin-sensitivity at really low Antimycin concentrations?

Low: Yes.

Sanadi: In general, when the Antimycin sensitivity is low, I would presume that the electrons are entering at the cytochrome c region. The differences, if any, may be due to the fact that the particles used in our respective laboratories are somewhat different. The preparative procedures do vary.

Low: Yes, that is our opinion, too.

Hommes: You had a rather large difference in Antimycin sensitivity between your succinate-driven reaction and your ascorbate-Q_1-driven reaction.

Sanadi: We get preparations with requirements ranging all the way from roughly 2 µg/mg protein to about 20 µg/mg protein, for 50 % inhibition. It varies with the preparation of the particles.

Hommes: But this, I assume, was done on the same preparation.

Sanadi: This was the identical preparation. This curve looks parallel, but it has been shifted, if you notice that. I do not know the significance of this.

Conover: To return to Professor Slater's last statement: do I understand that Low's Q-driven reduction of DPN and Sanadi's both had a low Antimycin sensitivity?

Chance: There is a real difference in sensitivities. Sanadi's is highly insensitive to Antimycin, and Low's is highly sensitive, under conditions of similar electron donors.

Estabrook: With respect to the spectrum, why did the reoxidized form not return to the original?

Sanadi: This is puzzling us. In the flavin region it returns but in the 408 mµ region it does not return.

Estabrook: Do you presume that there is some iron which becomes reduced, which is then not oxidized?

Sanadi: Not freely reoxidized by oxygen, perhaps, if the 408 mµ absorption has to do with iron. In that experiment we allow sufficient time for all of the added DPNH to be oxidized by following the 340 mµ absorption, so there was no more reducing substrate present. We don't know what would happen if the quinone is used as the oxidant, instead of oxygen.

THE ROLE OF PYRIDINE NUCLEOTIDE IN ENERGY LINKED REACTIONS
F. A. Hommes
Johnson Research Foundation, University of Pennsylvania
Philadelphia, Pennsylvania

The subtitle of this part of the colloquium is "Generation of Reducing Power" and we may ask if pyridine nucleotide itself is part of the mechanism of generation and utilization of reducing power. A detailed study of the succinate-linked DPN reduction in submitochondrial particles, which do not have endogenous DPN, in particular the influence of various DPN concentrations on the kinetic pattern of the reversal reaction, may give an answer to this question.

The reaction is shown in Fig. 1. To the particles suspended in buffer at high Mg^{2+} concentration are added sulfide as a terminal inhibitor, succinate as electron donor and ATP as energy source. The reaction is initiated by the addition of DPN and the formation of DPNH is followed fluorometrically. After about 2.5 min no further increase in fluorescence is observed. A further addition of ATP does not give rise to an increased formation of DPNH. That there is still DPN in the solution can be demonstrated by the addition of ethanol and yeast alcohol dehydrogenase which does give rise to an increase in fluorescence to a complete recovery of DPNH from the added DPN.

When this experiment was repeated at different DPN concentrations it became apparent that a rather constant amount of DPN was not reducible by succinate and ATP. This relationship is shown in Fig. 2, in which the amount of added DPN is plotted versus the amount of DPNH generated during the reaction. A slope of almost 1 is obtained at the higher DPN concentrations used. A marked deviation, however, is apparent at

---

Fulbright Scholar, 1961-1963. Permanent address: Department of Biochemistry, School of Medicine, University of Nijmegen, The Netherlands.

the lower DPN concentrations. In this particular preparation the equivalent of 0.7 mµ moles of DPN per mg particle protein is not reducible by succinate and ATP.

These experiments suggest that in the succinate-linked DPN reduction some form of DPN is an intermediate. A determination of the order of this reaction with respect to DPN gives a further support to this assumption.

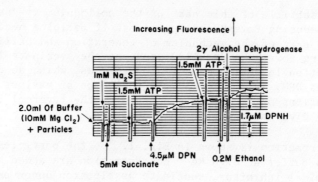

Fig. 1. Kinetics of the succinate-linked DPN reduction at low DPN concentration. Sonic particles prepared from beef heart mitochondria (1) were diluted in buffer (80 mM K Cl, 10 mM tri ethanol amine hydrochloride, 10 mM Mg Cl_2, pH 7.5) to a final protein concentration of 1 mg/ml. Sulfide was added as a terminal inhibitor, succinate as electron donor, followed by ATP. The reaction was started by the addition of DPN. The system was preincubated with ATP to abolish the lag phase which is usually observed in this reaction (2). DPNH formation was measured fluorometrically as described earlier (3).

The velocity of any reaction can be described by the formula $v = k \cdot c^n$, in which k is a reaction constant for a situation in which one of the reactants (c) of the reaction system in association with its enzymes is rate limiting. A log velocity versus log c plot gives then the order of the reaction (n) with respect to the reactant c. By applying this to the rates at the various DPN concentrations used in Fig. 2, an order with respect to DPN of 1.6 was found at low DPN concentration and of 0.4 at the higher DPN concentrations. This is shown in Fig. 3.

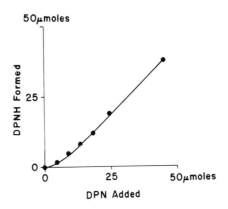

Fig. 2. Relation between the amount of DPN added to the reaction system and the amount of DPNH formed in the succinate linked DPN reduction. Experimental conditions as in Fig. 1.

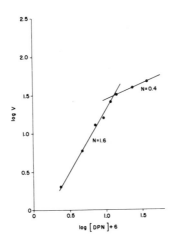

Fig. 3. Determination of the order with respect to DPN of the succinate linked DPN reduction from a double logarithmic plot of velocity versus initial DPN concentration. Experimental conditions as in Fig. 1.

The observation of an order higher than 1 indicates that DPN enters the sequence of reactions from succinate and ATP to DPNH at more than one site. The higher DPN concentrations saturate one site with the result that a pseudo lower order is found.

Any specific treatment of the reaction system which makes this site not rate limiting any more will result in an observed pseudo lower order with respect to DPN. We have found two examples of such a specific treatment.

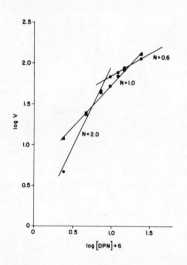

Fig. 4. Determination of the order with respect to DPN of the succinate linked DPN reduction from a double logarithmic plot of velocity versus initial DPN concentration, in the absence of a soluble mitochondrial factor (●—●) and in the presence of this factor (■—■).

First, the addition of a soluble mitochondrial factor which stimulates the rate of the succinate linked DPN reduction (1). Addition of this factor to the system decreased the order with respect to DPN from 2.0 and 0.6 to 1.0 when assayed over the same range of DPN concentrations. This is shown in Fig. 4.

Another example is pretreatment of the particles with the components of the reaction system, that is, sulfide, succinate, ATP and DPN. The particles were preincubated with these components for 5 min after which the particles were resuspended in their original volume of 0.25 M sucrose. The order with respect to DPN was then determined over the same range of DPN concentrations as with the untreated particles. The results are shown in Table I.

INFLUENCE OF PRETREATMENT OF SUBMITOCHONDRIAL PARTICLES ON THE ORDER WITH RESPECT TO DPN OF THE SUCCINATE LINKED DPN REDUCTION

Pretreatment	Order with Respect to DPN	
	Low DPN	High DPN
None	1.8	0.9
S⁻, succ., ATP, DPN		0.7
S⁻, succ., DPN	2.0	0.7
S⁻, DPN, ATP		0.7
S⁻, succ., ATP	1.3	0.7

Pretreatment with the complete system decreased the order of the reaction. As is shown, for this decrease in order both ATP and DPN are required, but not succinate. These are, however, conditions which may give rise to the formation of DPNH and therefore the question arises if DPNH itself can give this effect. Addition of DPNH prior to the addition of DPN gives, indeed, a decrease in order with respect to DPN when assayed under identical conditions over the same range of DPN concentrations. This is shown in Fig. 5. An order of 1.3 was found in the absence of added DPNH; in the presence of added DPNH the order with respect to DPN was 0.4. The minimum amount of DPNH required to initiate this transition was found to be 0.6 mµ moles of DPNH per mg particle protein, which is of the same order of magnitude as the amount of DPN which is not reducible by succinate and ATP.

All these experiments suggest that in the succinate-linked DPN reduction a DPN or DPNH containing compound functions as an intermediate in the reaction.

We may ask if such an intermediate is functioning in other pyridine nucleotide and energy linked reactions. The transhydrogenase reaction as described by Danielson and Ernster (4) is an example of such a reaction. This is shown in Fig. 6.

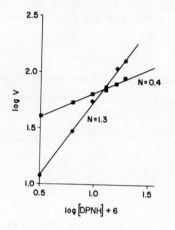

Fig. 5. Double logarithmic plot of velocity versus initial DPN concentration at various DPN concentrations with (■—■) and without pretreatment with DPNH (●—●).

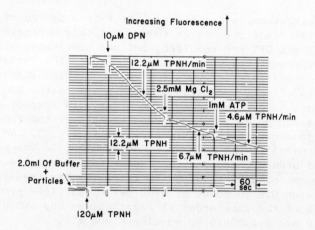

Fig. 6. Demonstration of the inhibition by Mg^{++} of the transhydrogenase reaction in sonic particles derived from beef heart mitochondria. TPNH is added to the particles suspended in buffer. TPNH is not oxidized until DPN is added. Any DPNH formed is oxidized by the submitochondrial particles. The decrease in fluorescence represents the oxidation of TPNH.

In this experiment TPNH is added to the particles suspended in buffer. The TPNH is not oxidized by the particles unless DPN is added. This oxidation is inhibited by Mg^{2+} and further inhibited by ATP. The reason for the inhibition by ATP will become apparent from the contribution by Dr. Estabrook. We will concentrate here on the inhibition by Mg^{2+}.

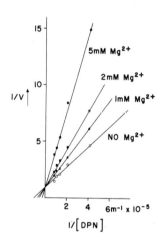

Fig. 7. Lineweaver-Burk plot of the transhydrogenase reaction TPNH + DPN $\longrightarrow$ TPN + DPNH at various DPN concentrations, demonstrating the competitive inhibition by Mg^{++}. Experimental conditions as in Fig. 6.

This metal ion proved to be a competitive inhibitor with DPN, as is shown in Fig. 7, with a K_i of 2.5 mM. It should be remembered that the succinate linked DPN reduction has an absolute requirement for Mg^{2+}. Maximum rates are observed at a Mg^{2+} to ATP ratio which exceeds by far the ratio expected if the only function of the Mg^{2+} was to facilitate the reaction with ATP. The apparent K_m of Mg^{2+} was found to be 3 mM. It should also be remembered that the previously mentioned soluble factor only exerts its rate stimulating capacity at low Mg^{2+} concentration (1). When added in excess and at high Mg^{2+} concentration it inhibits the reaction, which is shown in Fig. 8. The open circles are the rates in the absence of added factor as a function of the Mg^{2+} concentration, the triangles in the presence of 5 units of the factor, the crosses, closed squares, and closed circles in the presence of 10, 20, and 30 units of the factor respectively.

45

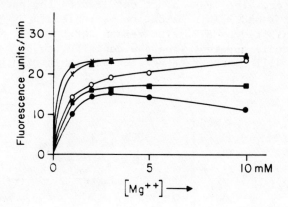

Fig. 8. Relation between the rate of DPN reduction measured in arbitrary fluorescence units and the Mg^{++} concentration, at different concentrations of mitochondrial factor. ▲:5 units added, x 10 units added, ■ 20 units added, ● 30 units added.

In the reverse of the above-mentioned transhydrogenase reaction, Mg^{2+} shows a mixed type of inhibition with DPNH. This is shown in Fig. 9. The way these experiments were carried out was, however, such that the DPN concentration was also varied, and it can be demonstrated that DPN is a competitive inhibitor with TPN. It looks, therefore, as if Mg^{2+} is a noncompetitive inhibitor with DPNH in this reaction.

The same titer of inhibition by triodothyronine and the unusual high activation energy of the succinate linked DPN reduction and the transhydrogenase reaction under various conditions have led us to believe that these reactions have a common intermediate. The inhibition studies with Mg^{2+} demonstrate that this metal ion is part of this intermediate.

A scheme is presented in Fig. 10 in which a DPN or DPNH containing intermediate somehow influenced by Mg^{2+}, has a central position for the transhydrogenase reactions and the succinate linked DPN reduction. It offers an explanation for the second order kinetics with respect to DPN in the succinate linked DPN reduction, the inhibition by Mg^{2+} of the transhydrogenase reactions and the release of this inhibition or increase of inhibition by ATP, and the absolute requirement of DPN for the succinate linked TPN reduction.

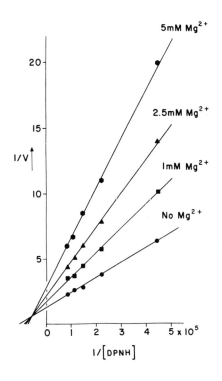

Fig. 9. Lineweaver-Burk plot of the transhydrogenase reaction DPNH + TPN $\longrightarrow$ DPN + TPNH at various DPNH and Mg^{++} concentrations. To the particles suspended in buffer were added sodium sulfide (1 mM), ethanol (60 mM), alcohol dehydrogenase (4 γ) and DPN. After the alcohol dehydrogenase reaction had reached equilibrium, TPN was added to start the transhydrogenase reaction. The DPNH concentration was varied by adding various amounts of DPN to the alcohol dehydrogenase system.

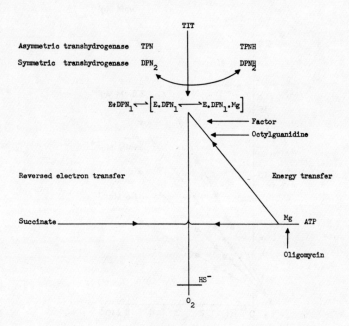

Fig. 10. Schematic representation of the transhydrogenase reaction and the succinate linked DPN reduction in submito-Chondrial particles. For explanation see the text.

REFERENCES

1. Hommes, F. A., Biochim. Biophys. Acta, 71, 595 (1963).
2. Hommes, F. A., Biochim. Biophys. Acta, in press.
3. Hommes, F. A., Biochem. Biophys. Res. Comm., 8, 248 (1962)
4. Danielson, L., and Ernster, L., Biochem. Biophys. Res. Com 10, 91 (1963).

DISCUSSION

<u>Griffiths</u>: I was interested in your suggestion as to the form of DPN which acts in this reaction. This quantity of DPN which does not react—do you have any evidence, other than that afforded by the reaction with ADH, that this is not NAD?

<u>Hommes</u>: It reacts with alcohol and alcohol dehydrogenase. This is the only evidence I have. I don't know whether it is related to extra DPN.

<u>Slater</u>: I was very much interested in that observation, and in your comments on the residual DPN at the end of this experiment which could not be reduced by succinate + ATP, but was reduced by alcohol and alcohol dehydrogenase. As I was planning to say this afternoon, we have found in the alkaline extract of a very similar reaction mixture an alkali-extractable DPN which is reduced by alcohol and alcohol dehydrogenase but which cannot be DPN because DPN cannot stand up to that alkaline extract treatment. This is the same type of pyridine nucleotide compound which Hilvers found earlier in the glyceraldehyde phosphate dehydrogenase reaction. The alkali-extractable compound was assayed as DPN but could not be DPN because DPN is unstable in the alkaline extract. Yours now appears to be the same type of compound as in the alkaline extract of the reaction mixture. Although we haven't done your experiment, it looks as if they agree with one another very well, indeed.

COUPLING OF ENERGY LINKED FUNCTIONS
IN MITOCHONDRIA AND CHLOROPLASTS TO THE
CONTROL OF MEMBRANE STRUCTURE
Lester Packer, Reginald H. Marchant and Yasuo Mukohata
Department of Physiology, University of California
Berkeley, California

The requirement of biological reductions in aerobic systems for an energy source and reducing power is being studied in mammalian heart mitochondria using a terminal phosphorylating shunt of the respiratory chain. Our interest in this problem originated from certain experiments on the mechanism whereby the respiratory chain phosphorylating process drives rapid and reversible swelling-shrinking changes which accompany changes in the metabolic state of mitochondria (1-3). During the course of these studies we noted that electron transport through restricted segments of the respiratory chain were capable of driving reversible changes in mitochondrial volume (3). In particular ascorbate - tetramethylphenylene diamine, which catalyzes a cytochrome oxidase feeder system and which was discovered by Dr. E. E. Jacobs (4), was found most effective in this regard and we have reported on this in the last two years (5-7). These observations prompted a closer look at its role in the mitochondrial electron transport system with a view to determining its usefulness as a tool in studying the respiratory enzymes as a chemi-osmotic transducer.

I would like to begin by illustrating some of the characteristics of the ascorbate-TMPD feeder system. Figure 1 shows oxidative phosphorylation and respiratory control associated with ascorbate-TMPD oxidation. The experiment was begun by adding mitochondria to the basic reaction mixture; this decreased the dissolved O_2 concentration of the system by dilution. Antimycin A was then added to ensure no oxidation of endogenous substrates. Following this the feeder substrate, ascorbate-TMPD, was added. Several additions of TMPD were made to raise the respiratory rate to 0.79 µM/second. Addition of ADP accelerated respiration 1.84 times; this acceleration lasted for about one minute, after which it slowed. The accelerated respiration corresponded to the period of phosphorylation, and during this period, a total of 82 µM O_2 (or 164 µatoms O/liter) was consumed. This corresponds to an ADP:O of 0.99, in excellent agreement with our earlier findings.

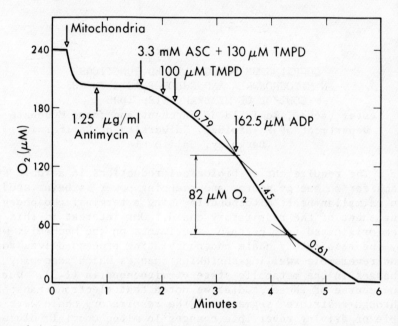

Fig. 1. Respiratory control associated with the terminal phosphorylating site of the mitochondrial respiratory chain. Rat liver mitochondria were prepared in 0.25 M sucrose as previously reported (4), and respiration and phosphorylation were assayed polarographically with the Clark oxygen electrode. The polyethylene coated Clark electrode was of considerable advantage for these studies because TMPD reacts somewhat with uncoated electrodes. The reaction system contained: sucrose (0.067 M), KCl (0.02 M), $MgCl_2$ (0.005M), phosphate (0.02 M, pH 7.5), and mitochondria (0.85 mg protein/ml). The additions are indicated in the figures and text. Temperature was maintained at $25° \pm 0.1°$ C. The reaction mixture was mixed with a magnetic stirring device.

The action of oligomycin on the ascorbate-TMPD system is illustrated in Fig. 2. The oligomycin was introduced into the reaction system after antimycin. The subsequent addition of the feeder substrate (ascorbate-TMPD) produced an O_2 consumption of 0.85 μM/second. Introduction of ADP into the reaction mixture left the respiratory rate unchanged, thus

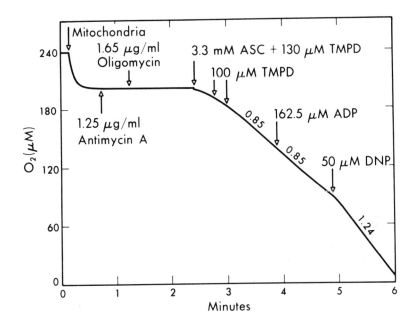

Fig. 2. Action of oligomycin on respiratory control associated with the terminal phosphorylation site of the mitochondrial respiratory chain. Conditions as in Fig. 1.

showing loss of respiratory control. Release of the inhibition was brought about by 50 μM DNP, in agreement with the action of uncoupling agents of this type at other oligomycin-inhibited phosphorylation sites in the respiratory chain. Concentrations of oligomycin less than 1.6 μg/ml also abolish the phosphorylating respiration in the terminal part of the chain, but longer periods of incubation are required as the concentration of oligomycin is decreased. It may be noted that the low respiratory control index of about 2, using ascorbate-TMPD, results in an appreciable residual respiration in the presence of oligomycin.

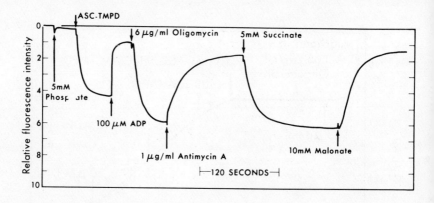

Fig. 3. Control of electron and energy transfer from the terminal cytochrome region of the mitochondrial respiratory chain. Rabbit heart mitochondria were isolated in sucrose (0.32 M) - ethylene diamine tetraacetate (0.001 M, pH 7.0) as previously reported (4). Pyridine nucleotide fluorescence was excited at 365 mµ and monitored at 450 mµ, using a photomultiplier positioned at 90° to the exciting light in a Brice Phoenix light-scattering photometer modified for recording (3). The basic reaction system contained: sucrose (0.07 M), KCl (0.02 M), Tris-hydroxy-methyl aminomethane buffer (0.02 M pH 7.5), and mitochondria (0.68 mg protein/ml). The concentration of the feeder substrate was 0.83 mM ascorbate + 134 µM TMPD. See text for explanation. NAD reduction is registered as an increase in fluorescence intensity (8). The fluorescence intensities are corrected for the background intensity of the mitochondrial suspension.

The manner in which the terminal cytochrome region can control some of electron and energy transfer functions of mitochondria can be illustrated by coupling the terminal section of the respiratory apparatus to the generation of reducing power; that is observing the oxidation-reduction state of intramitochondrial pyridine nucleotide. This experiment is illustrated in Fig. 3, which is a time recording of the NADH fluorescence intensity changes shown by rabbit heart mitochondria. It might be mentioned that rabbit heart mitochondria show the same antimycin insensitive and oligomycin sensitive terminal phosphorylation as liver mitochondria. The experiment was commenced by addition of phosphate to the reaction system; this did not alter the fluorescence intensity of the mitochondrial suspension, indicating a lack of endogenously reduced pyridine nucleotide. When endogenous substrate is present, phosphate promotes its disappearance, and an oxidation of NADH would be anticipated; this was not observed. Endogenous substrates were readily removed from cardiac mitochondria by repeated washing in appropriate isolation media, a necessary condition for studying the dependence of NAD reduction under the experimental conditions.

The next step in the experiment was to initiate electron flow in the terminal feeder system with ascorbate-TMPD. This led to a rapid NAD reduction, thus indicating that both the energy and the reducing requirements for reversed electron transport were being generated from the terminal region of the chain (if antimycin was absent). Reoxidation of the NADH can be effected by coupling energy production in the terminal region to ATP synthesis through the addition of ADP. Energy transfer is apparently used preferentially for ADP phosphorylation rather than for reversed electron flow and the NADH is reoxidized. By blocking ATP synthesis with oligomycin, "trapped" energy in the terminal region was once again available for the reduction of mitochondrial NAD. As shown in Fig. 3, this was promptly brought about by 6 µg/ml of oligomycin. Since the terminal region supplies both the energy and the reducing power for NAD reduction, reoxidation of the NADH should be induced by cutting off the source of reducing power from the terminal region using antimycin A. It has been previously noted that the antimycin block "leaks" or that there is a bypass. Thus 1 µg/ml of antimycin A promoted the reoxidation of the NADH over a two minute interval. Further proof that this interpretation is correct is given by the observation that succinate, whose oxidation through the cytochrome region was now blocked by antimycin, again reduced the NAD by providing reducing power. The reducing power of succinate was abolished by its competitive inhibitor malonate, which once more caused re-oxidation of the pool of intramitochondrial NADH in the aerobic steady state. These results confirm the recent findings of Tager, Howland and Slater (9), who have also reported inhibition of phosphorylation with oligomycin in a complex system in which mitochondria were oxidizing TMPD without respiratory control. Tager and co-workers (9) and Penefsky (10) have also found that oligomycin does not inhibit a TMPD linked reduction of α-ketoglutarate plus NH_3 (to glutamate), or of ubiquinone, respectively.

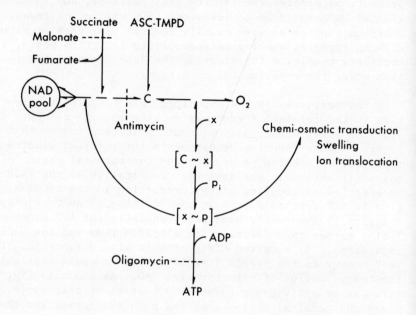

Fig. 4. Scheme illustrating proposed energy coupling of the ascorbate-TMPD feeder system to mitochondrial swelling, pyridine nucleotide reduction, and ion translocation.

Fig. 4 sketches our present view of the pathways involved. This rather complex experiment (Fig. 3) made with two exogenous sources of reducing power and three selective inhibitors clearly indicates utilization of energy generated in the terminal region of the respiratory chain for the control of oxidation-reduction changes in mitochondria, at a site in the respiratory system much earlier in the multi-enzyme sequence. It suggests that some fluid mechanism may exist for the transfer of energy to spatially separated regions of the system. The action of oligomycin clearly shows that this energy transfer function is independent of ATP itself.

We next turned our attention to taking a closer look at the role of the ascorbate-TMPD feeder system as an energy source of mitochondrial volume changes, and in general to explore its usefulness for studying mitochondria as chemiosmotic transducers.

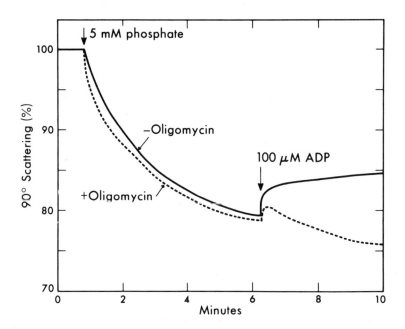

Fig. 5. Swelling-shrinkage changes associated with electron flow through the ascorbate-TMPD feeder system. The reaction mixture at 25° C contained Tris buffer (0.02 M, pH 7.5), KCl (0.02 M), sucrose (0.05 M), ASC (0.83 mM), TMPD (133 μM), and rabbit heart mitochondria (0.06 mg protein/ml). Light-scattering changes at 90° were measured at 546 mμ in a Brice-Phoenix Light-Scattering photometer. The initial scattering intensity of the mitochondria in the basic reaction mixture was taken as 100%. The decreases in scattering (swelling) and scattering increases (shrinkage) were observed over the time period indicated in the absence and presence of oligomycin (4 μg/ml).

The characteristics of the swelling-shrinkage changes associated with electron flow through the ascorbate-TMPD system are illustrated in Fig. 5. It may be seen that the addition of phosphate promotes a rapid decrease in scattering (swelling) reaching completion in approximately 6 minutes. The time course of the swelling change promoted by ascorbate-TMPD + phosphate (both of which are required) is virtually the same in the presence or absence of oligomycin. The introduction of ADP into the system which triggers off oxidative phosphorylation leads to a small degree of shrinkage in the control incubated in the absence of oligomycin. It can be seen that this partial reversal of the swelling change brought about by ADP is not observed in the presence of oligomycin; this is to be expected since oligomycin abolishes the respiratory control and ATP synthesis associated with the ascorbate-TMPD system.

The scheme shown in Fig. 4 indicates that the accumulation of the phosphorylated intermediate, $X \sim P$, would not be inhibited in the presence of oligomycin. The correlation of $X \sim P$ with the swelling response is suggested by other considerations. It is generally observed with mitochondria that the state of least swelling is obtained with substrate only ($X \sim P$ concentration minimal); the maximum swelling state is brought about by substrate + phosphate ($X \sim P$ concentration maximal); the intermediate state of swelling lying between these two extreme cases is obtained with substrate + phosphate + ADP ($X \sim P$ concentration reduced by ATP synthesis). Indeed, the experiment just discussed showed these three levels of swelling for the ascorbate-TMPD system.

The scheme in figure 4 indicates that accumulation of $X \sim P$ would not be inhibited in the presence of oligomycin. Currently, however, there is some uncertainty as to the exact site of action of oligomycin. If oligomycin acts prior to $X \sim P$, then any energy requiring process which still proceeds in its presence must be capable of utilizing some earlier intermediate than $X \sim P$ as a source of energy, and must also be independent of phosphate. Mitochondrial swelling is independent of the oligomycin block and dependent on phosphate; thus either the site of oligomycin action indicated in figure 4 is correct, or the role of phosphate in this phenomenon is completely independent of its role in $X \sim P$ formation. Since mitochondrial shrinkage has normally been regarded as an energy requiring process, the dependence of swelling on substrate and phosphate is in a sense contradictory. Either swelling is an energy requiring process, or as an alternative interpretation of these results, some earlier intermediate than $X \sim P$ promotes shrinkage and the formation of $X \sim P$ in the presence of phosphate and substrate causes swelling by depleting this earlier intermediate necessary for the state of greater shrinkage.

It is tempting to think that some of the mitochondrial coupling factors with ATPase activity, including the recently described actomyosin-like protein isolated from mitochondria by Ohnishi and Ohnishi (11), may be directly involved in the control of these plastic properties of mitochondria. We have confirmed Ohnishi's results, and I understand that Professor Lehninger and his associates have as well. We show here four possible fates of intermediates as $X \sim P$. It may be used for 1) driving mitochondrial swelling, 2) ion translocation, a process which is also oligomycin insensitive as has been recently reported by the fellows in Wisconsin, 3) reversed electron transport with reduction of pyridine nucleotide, and 4) be conserved as ATP.

It seemed of interest to examine the magnitude of the swelling-shrinkage changes associated with the ascorbate-TMPD feeder system under conditions where some of the reducing power and energy were being competed for by the reversed electron flow mechanism. Table I summarizes several experiments with the ascorbate-TMPD system in the presence and absence of antimycin A or oligomycin. The actions of oligomycin on the swelling-shrinkage changes are similar to those already mentioned. It might be worth mentioning further, that the amplitude of the swelling is generally slightly greater in the presence of oligomycin. In the presence of antimycin A, ascorbate-TMPD induced a swelling response which according to the hypothesis presented here results from the production of $X \sim P$ as a consequence of electron flow in the terminal region. This swelling is almost completely reversed by the addition of ADP which would reduce the $X \sim P$ concentration. In the absence of antimycin A, the swelling is greater, and this must be due to some extra contribution made by the reversed electron transport which now occurs. This higher degree of swelling is not so completely reversed by the addition of ADP. Perhaps this result may be interpreted to indicate that altered levels of the energy containing compounds associated with the carriers in the early part of the respiratory chain are able to induce swelling, but ADP apparently cannot modify these levels to an appreciable extent. The competition for oxidized cytochrome c by TMPD and reduced pyridine nucleotide would appear to offer a plausible explanation of this result.

TABLE I

Swelling-Shrinkage Changes Associated With The Ascorbate-TMPD Feeder System

Reaction conditions: sucrose (0.05 M), KCl (0.02M), Tris buffer (0.02 M, pH 7.5), ascorbate (0.8 mM), TMPD (133 μM), mitochondria (0.64 mg protein/ml). Where indicated, the other concentrations were: phosphate (5 mM, pH 7.5), ADP (100 μM), antimycin A (1 μg/ml), and oligomycin (4 μg/ml). ( ) = number of experiments for which data were averaged.

Conditions	% Change in Light-Scattering		% Reversal of Swelling
	Swelling with P_i	Shrinkage with ADP	
No Addition	-16.6 (6)	+4.4 (5)	27
Antimycin A	- 9.0 (5)	+8.0 (9)	89
Oligomycin	-20.5 (5)	-0.4 (5)	0

I might summarize these results by saying that we feel that under conditions where the synthesis of ATP is prevented by oligomycin, other energy requiring processes such as reversed electron transport, swelling, and ion translocation, are unaffected, and this seems to dictate a compound such as X∼P to be functional in regulating the energy available for these processes.

Now we reasoned, that if such structural changes are a general phenomenon of energy transducing systems in nature that they should also be manifest in analogous systems, in particular those associated with the chloroplast membrane, which catalyze phosphorylation linked to photochemical reactions.

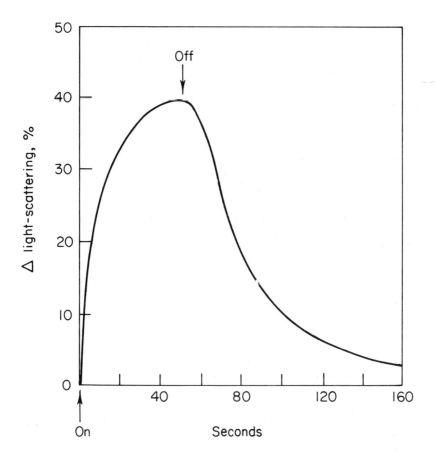

Fig. 6. Reversible scattering changes induced by red light illumination in spinach chloroplasts. The reaction system contained Tris (0.02 M, pH 7.7), NaCl (0.035 M), $MgCl_2$ (0.005 M), ADP (1 mM), phosphate (0.004 M), ascorbate (2.5 mM) PMS (20 μM) and chloroplasts (2 μg/ml chlorophyll). Explanation in text.

To put this notion to test, spinach chloroplasts were isolated in 0.35 M NaCl or 0.5 M sucrose buffered with Tris at pH 8.0. The chloroplasts were then incubated in the presence of the complete requirements for electron transport and photophosphorylation. The reaction was carried out in a 1 x 1 cm cuvette in a Brice-Phoenix Light-Scattering photometer modified for recording. The light-scattering changes of the chloroplasts were measured at $90°$, using incident light of 546 mμ. The scattered light was filtered at the same wave length (546 mμ) with an interference filter, and the scattering intensity was adjusted to read 100% on the chart paper using the minimum intensity of 546 mμ incident light, and the instrument at maximum gain. Also since light in this region is near the minimum of the photochemical action spectrum, this procedure minimizes the possibility of electron flow being activated by the incident beam. Increases and decreases in scattered light intensity in response to actinic red light were then expressed as a percentage of the initial level. The temperature was accurately controlled during periods of illumination at $25° \pm 0.1°$ C, by circulating liquid around the jacketed cuvette.

Fig. 6 shows the results of a typical experiment. Chloroplasts were incubated under conditions necessary for phosphorylation, that is, in the presence of actinic light, an electron carrier such as PMS, phosphate, ADP and Mg^{++}. A low concentration of chlorophyll, usually less than 10 μg/ml was employed to prevent pigments from interfering with the light-scattering measurements. It can be seen that on illuminating chloroplasts with red light, a rapid increase in scattering is observed, which reaches a steady state approximately 60 seconds after illumination. The increased scattering in the steady state was 40% greater than prior to red light illumination. Upon removal of the actinic light, the light-scattering by the chloroplasts returns to the initial level.

It was shown in collateral experiments that these optical changes were dependent on the angle at which the emitted green light was measured in a manner characteristic of scattering and that they are therefore not fluorescence changes. Absorbance changes were ruled out by showing that the % change was independent of the chloroplast concentration in the range used in the experiments, as well as by the fact that the absorbance of the pigments was too low to effect an apparent scattering measurement.

Fig. 7 shows an analysis of the growth and decay of the scattering response shown earlier. The half time, or time for a 50% scattering change was 6.6 seconds for the growth, and 22 seconds for the decay. It can be seen that the rise and decay are exponential. We do not know whether these first order kinetics are dependent upon a single chemical change or on a complex chain of reactions which give apparent first order kinetics.

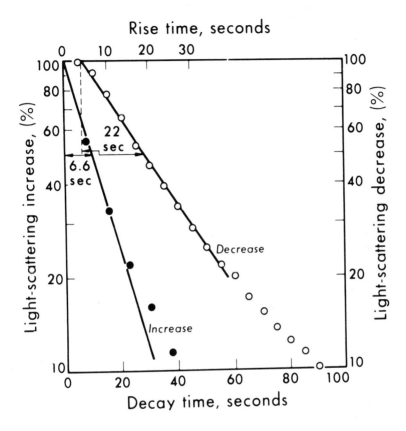

Fig. 7. Kinetic analysis of the growth and decay of scattering changes in spinach chloroplasts. Conditions as in Fig. 6.

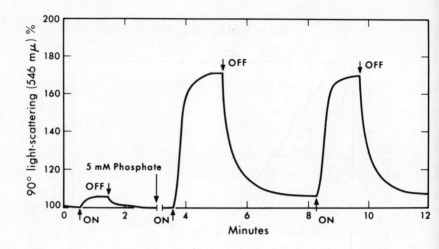

Fig. 8. Effect of phosphate on scattering changes induced by red light illumination in spinach chloroplasts. Conditions as in Fig. 6, but without phosphate.

Fig. 8 shows an example of the dependence of the scattering response on one of the requirements for photophosphorylation demonstrated by withdrawing phosphate from the complete system. In the absence of phosphate, a scattering response to red light of approximately 5% was obtained which was reversible in the dark. 5 mM phosphate was then added in the dark; upon illumination with actinic light, an enormously enhanced scattering response was obtained, and now it is seen that the scattering increase was about 70% above the initial level. It should be further noted that these responses are repeatable; alternating periods of light and darkness continue to give the scattering responses.

Fig. 9 shows a scheme for electron transport and photophosphorylation taken from one of Professor Arnon's recent articles (12). It illustrates the point at which various inhibitors are thought to act. If scattering responses are being studied in non-cyclic systems, using either TPN or ferricyanide as electron acceptors, the red light induced scattering responses are completely abolished by 10^{-6} M 3-(3,4-dichlorophenyl)-1, 1-dimethylurea (DCMU) which blocks the first light reaction. The ability to manifest these scattering changes can be restored to the system by adding ascorbate and a trace of the dye, 2,6 dichlorophenol indophenol (DCPIP), which

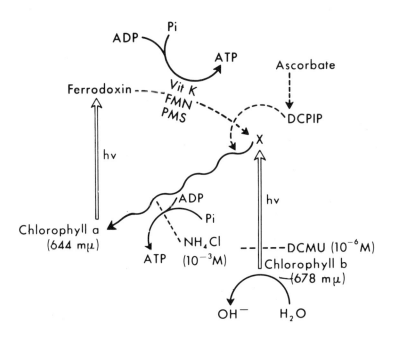

Fig. 9. Scheme for photosynthetic electron transport and cyclic phosphorylation in spinach chloroplasts, after Arnon (12).

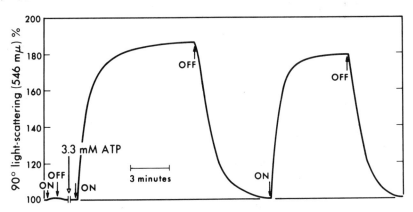

Fig. 10. Restoration of reversible scattering changes in aged chloroplasts by ATP. The reaction system contained Tris (0.02 M, pH8.0), NaCl (0.035 M), $MgCl_2$ (0.005 M), phosphate (0.004 M pH 8.0), ADP (1 mM), ascorbate (2.5 mM), 2,6-dichlorophenol indophenol (30 μM), phenazine methosulfate (20 μM), and chloroplasts (approximately 10 μg chlorophyll/ml).

reinstitutes electron flow and phosphorylation by by-passing the DCMU block. The scattering changes under these conditions can again be abolished by 1 mM NH_4^+ or by one of a number of uncoupling agents such as m-chloro-carbonyl cyanide phenylhydrazone (CCP) or pentachlorophenol (PCP). In separate experiments, it can also be shown that cyclic photophosphorylation initiated with either phenazine methosulfate, flavin mononucleotide, or vitamin K_3 also permits reversible scattering responses. The magnitude of the scattering response also varies with the intensity of the actinic light, although a careful study of this relationship for the several systems has not yet been made. In short, all of the known effects of actinic light, electron carriers, phosphate acceptors, and inhibitor substances in the chloroplast system seem to promote or prevent scattering responses in an exactly predictable manner.

Although these studies established that light-scattering increases in chloroplasts occur under conditions of cyclic and non-cyclic photophosphorylation, it was noted in some cases that scattering changes were lost even when chloroplasts were incubated in the presence of complete requirements for photophosphorylation. These "aged" chloroplasts were found to require addition of ATP in order to manifest the scattering changes. Fig. 10 dramatically demonstrates this ATP requirement. Illumination of chloroplasts under conditions for cyclic photophosphorylation led to an increase in scattering of 2% upon illumination with red light. This small scattering increase was reversed after extinguishing the actinic light. The addition of 3.3 mM ATP in the "dark" did not change the scattering level. However, when the red light was restored, a rapid and extensive increase in scattering ensued, reaching a steady state at a level of 85% higher than the initial scattering intensity. This increased scattering state could be fully reversed by extinguishing the actinic light. A second light and dark period led to a similar cycle of scattering increase and decrease. Thus the large reversible scattering responses, characteristic of "fresh" chloroplasts, had been restored by ATP. A similar improvement of the scattering response can be obtained with ITP.

This remarkable restoration of the scattering response induced by red light in the presence of ATP suggested that this action of ATP might bear some relation to the existence of light-induced ATPase in chloroplasts which have been reported on by Avron (13) and Petrack and Lipmann (14).

TABLE II
Correlation of Photohydrolysis of ATP and ITP by Spinach Chloroplasts With Changes in Chloroplast Structure

The reaction mixture contained Tris buffer (0.02 M, pH 8.0), NaCl (0.035 M), $MgCl_2$ (0.005 M), ascorbate (0.83 mM), phenazine methosulfate (20 µM), and chloroplasts (12 µg chlorophyll/ml). The concentration of the nucleoside triphosphates were 2 mM, and that of cysteine, 0.083 M, where they appear in the table. The chloroplast preparation used in this experiment was prepared in the usual manner in Tris (0.1 M, pH 8.0) - NaCl (0.35 M), and then washed two additional times by centrifugation in a low salt medium, Tris buffer (0.005 M, pH 8.0) - NaCl (0.035 M). This procedure removes the soluble protein of chloroplasts. Illumination in the scattering experiment was with red light as described in methods. The extent of the % scattering increase is given by the (+) values and the extent of the decay in scattering after removal of actinic light by the (-) values. In parallel experiments made on nucleoside triphosphatase activity, illumination was made with an unfiltered Tungsten light source. The chlorophyll concentration in the photohydrolysis experiment was 200 µg chlorophyll/ml; other conditions were the same as in the scattering experiment.

Condition	% Scattering change		µmoles phosphate formed/mg chlorophyll/15 min.	
	Light	Dark	Light	Dark
ATP	+22	-18	0.37	0.38
ITP	+21	-21	0.10	0.22
cysteine + ATP	+69	-9	2.73	0.39
cysteine + ITP	+83	-21	1.11	0.12

Petrack and Lipmann demonstrated that light-induced ATPase of spinach chloroplasts was maximally activated in the presence of cysteine (0.08 M). Accordingly the action of ATP and also of ITP on the scattering responses in fresh chloroplasts was tested in the presence and absence of cysteine. Parallel determinations were made of nucleoside triphosphatase activity. The results are shown in Table II. They confirm the existence of the light activated ATPase observed by Petrack and Lipmann (14), and also show photohydrolysis of ITP. Under conditions where nucleoside triphosphatase activity is a maximum (i.e. in the presence of cysteine), it may be seen that the scattering increases induced by red light under conditions of cyclic photophosphorylation are considerably larger as compared with those in the absence of cysteine. The scattering changes observed in the absence of cysteine when the nucleoside triphosphatase activity is lower, are largely reversed in the dark, as illustrated above. However, in the presence of cysteine, the scattering increase is only slightly reversed when the red light is turned off. Addition of one of a number of substances which inhibit light-induced ATPase or ITPase activity, such as NH_4Cl (1 mM) or ADP (1 mM), bring about a full reversal of the scattering response. This is not shown here. These findings suggest an involvement of light-induced nucleoside triphosphatase activity in the control of structural changes geared to photosynthetic electron transport

TABLE III
Light-Scattering Changes and Nucleoside Triphosphatase
Activity of a Protein Extract of Chloroplast Membranes

Light-scattering changes and nucleoside triphosphatase activity were determined as described in the methods. The experiments at low ionic strengths were carried out in 0.04 M KCl - 0.12 M imidazole, and those at high ionic strengths in 0.6 M KCl - 0.12 M imidazole at the pH indicated. ATP and ITP concentrations were 4 mM.

	pH 5.5		pH 6.3		pH 7.5	
	ATP	ITP	ATP	ITP	ATP	ITP
Low Ionic Strength						
% Light-scattering decrease	16.0	16.0	8.0	6.5	7.0	6.5
Nucleoside triphosphatase*	2.10	1.70	1.50	1.40	0.25	0.21
High Ionic Strength						
% Light-scattering decrease	2.0	3.0	--	--	--	--
Nucleoside triphosphatase*	0.32	0.28	--	--	--	--

* μmoles phosphate formed/mg protein/30 minutes at $25°$ C.

A similarity between these results and the properties of the contractile protein of muscle was at once apparent. Accordingly chloroplasts were extracted with 0.6 M KCl under similar conditions to those employed for the extraction of actomyosin from mammalian muscle. Some of the properties of this extracted protein fraction are shown in Table III.

It may be seen that this protein fraction extracted from chloroplast membranes manifests ATPase and ITPase activity. The same fraction also shows decreases in light-scattering on addition of these nucleotides. Nucleoside triphosphatase activity is greater at pH 5.5 than at higher pH's. Likewise, the ability of these nucleotides to cause a decrease in scattering is also greater at the lower pH. The nucleoside triphosphatase activity at high ionic strength was lower than at low ionic strengths (pH 5.5). Similarly, the light-scattering responses observed in the preparations at high ionic strength were much smaller than those observed at low ionic strengths. Parallel experiments on the viscosity of the protein extract in the presence and absence of the nucleotides were also made. These results did not show a decrease in viscosity with added nucleotides, which would be expected if this preparation behaved like the contractile proteins of mammalian muscle and mitochondria.

Although it may be too early to conclude that chloroplasts contain a contractile protein, the conclusion that ATPase (or ITPase) activity is somehow involved in light-induced changes of chloroplast structure seems inescapable. This conclusion is based upon the following evidence: first, both ATP and ITP restore to aged chloroplasts the ability to manifest structural changes under conditions of photosynthetic electron transport. Secondly, under conditions where ATPase and ITPase activities are maximally activated in fresh chloroplasts (in the presence of cysteine), scattering increases with red light are larger and no longer reversible when red light is extinguished. But the reversal of the scattering changes may be completed by addition of known inhibitors of chloroplast nucleoside triphosphatase. Lastly, a protein fraction isolated from chloroplast membranes under similar conditions to those employed for extraction of contractile proteins from muscle shows ATPase and ITPase activity and decreases in scattering with ATP and ITP, which suggest the existence of conformational changes. These results are indicative of a primary role for ATP, the product of photophosphorylation, in the control of chloroplast structure. We think this approach may provide some insight into the nature of the intermediates in photophosphorylation and for the question of the reversal of this energy transfer pathway in chloroplasts.

Our present view is summarized in Fig. 11.

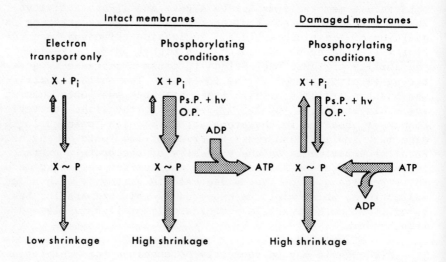

Fig. 11. Hypothesis for coupling of energy-transducing systems to membrane structure of mitochondria and chloroplasts.

It is hypothesized that in chloroplasts and mitochondria a coupling exists between the energy transducing system and their membrane structure. In the intact membranes of chloroplasts and mitochondria a low light-scattering and low-shrinkage state exists during electron transport in the absence of phosphorylation. Under conditions of oxidative phosphorylation and photophosphorylation, energy-linked intermediates are synthesized in increasing concentration. The fate of these intermediates is then two-fold; their energy can be conserved to make ATP or alternatively utilized for coupled mechanochemical work, or the translocation of ions. In damaged membranes, however, high concentrations of an energy linked intermediate generated by the phosphorylation process are difficult to maintain because of its hydrolysis under these conditions. Thus in both mitochondria and chloroplasts, it appears that ATP is required to maintain sufficient levels of the intermediate to drive structural changes. Activation of the light induced ATPase with cysteine allows the direction of the reactions to be shifted from ATP synthesis toward the accumulation of energy-linked intermediates in the fresh chloroplasts. So now we find that the amplitude of the scattering increase is larger when chloroplasts are incubated under these conditions, and in fact, if the concentration of ATP is high enough, the response is not readily reversed in the dark unless an ATPase inhibitor as NH_4Cl or ADP is added.

Recently Ito, Izawa and Shibata (personal communication) have observed that spinach chloroplasts undergo shrinkage and a change in axial ratio (from 1.96 to 2.34) on illumination. Since these changes are reversed in the dark they may be regarded as a possible indication of the nature of structural changes related to the phosphorylation process in chloroplast membranes.

In conclusion, whatever the final physiological significance of the physical change, and the precise mechanism of this process may be, it would seem important that the primary energy transducing systems of living cells, the mitochondrion and the chloroplast, are both capable of bringind about structural changes in their membranes in a predictable manner.

REFERENCES

1. Packer, L., J. Biol. Chem., 235, 242 (1960).
2. Packer, L., J. Biol. Chem., 236, 214 (1961).
3. Packer, L., J. Biol. Chem., 237, 1327 (1962).
4. Jacobs, E.E., Biochem. Biophys. Res. Comm., 3, 536 (1960).
5. Packer, L. and Jacobs, E.E., Biochim. Biophys. Acta, 57, 371 (1962).
6. Packer, L., Biochim. Biophys. Acta, in press.
7. Packer, L., Marchant, R.H., and Corriden, E., Biochim. Biophys. Acta, in press.
8. Estabrook, R.W., Anal. Biochem., 4, 231 (1962).
9. Tager, J.M., Howland, J.L. and Slater, E.C., Biochim. Biophys. Acta, 58, 616 (1962).
10. Penefsky, H.S., Biochim. Biophys. Acta, 58, 619 (1962).
11. Ohnishi, T. and Ohnishi, T., J. Biochem., 51, 380 (1962).
12. Arnon, D.I., Fed. Proc., 20, 1012 (1962).
13. Avron, M., J. Biol. Chem., 237, 2011 (1962).
14. Petrack, B. and Lipmann, R., in Light and Life, McElroy, W.D. and Glass, B. (Editors), Johns Hopkins Press, Baltimore, 1961, p. 621.
15. Ito, T., Izawa, S. and Shibata, K., personal communication.

DISCUSSION

Chance: To return to the energy-linked transfer: what was the effect of malonate on the fluorescence increase, if malonate were added prior to succinate and Antimycin; is succinate an electron donor in this reaction?

Packer: I don't recall having done that.

Chance: Is there also electron donation from the cytochrome c level?

Slater: Yes, from ascorbate and TMPD. We have done the exact experiment you asked and would confirm Packer's results absolutely. We measured DPNH directly in rabbit-heart mitochondria DPNH formation was linked to glutamate synthesis in liver mitochondria (see p. 99).

Hess: I would like to continue the question. What is the electron acceptor and what happens to the DPNH?

Packer: We can also show the extent to which the pyridine nucleotide is reduced in the steady state in this Antimycin-blocked system, with succinate on the one side and the energy-feeder system on the other. You suppress the level of the pyridine nucleotide reduced in the steady state in the presence of ascorbate by increasing the concentration of fumarate, indicating that the role of succinate is only as a source of electron donation.

Hess: You did not measure the ascorbate concentration. Is there any indication that dehydroascorbate is involved?

Davies: Does peroxide accumulate due to the ascorbate oxidation?

Packer: I doubt it, because the reaction starts rapidly.

Davies: Then the reaction is catalase-insensitive.

Packer: Yes, it is catalase-insensitive.

Slater: A small but important point of terminology which, as you linked my name with it, I would like to bring out. Oligomycin is certainly not an uncoupler of oxidative phosphorylation. From the work of Lardy (1) and all subsequent work on oligomycin, it is quite clear that oligomycin is an inhibitor of oxidative phosphorylation, not an uncoupler.

Packer: I agree.

Slater: The other point is also almost terminology, but it is important, too. I don't see how you can retain $X \sim P$ as being necessary for reversed electron transfer, when Ernster (2) showed, and this was confirmed by Snoswell (3), that the reaction can proceed without phosphate.

Packer: We could show a phosphate dependence for pyridine nucleotide reduction with succinate; in a well-washed mitochondrial system, the reaction also requires ATP or succinate.

Estabrook: In the presence of oligomycin, do you need phosphate in order to get the reduction of pyridine nucleotide?

Slater: The phosphate experiment of Ernster and Snoswell shows that you don't need $X \sim P$.

Chance: Phosphate inhibits DPN reduction in intact mitochondria.

Estabrook: Do you envisage a mobility or an immobilization of the high-energy intermediate? Are you implying that there is but one intermediate?

Packer: No.

Estabrook: Or that the intermediate diffuses from the cytochrome oxidase site to the pyridine nucleotide? There must be cross-communication between the two.

Packer: There must definitely be cross-communication. How this is accomplished in the spatially separating site, I don't know.

REFERENCES

1. Lardy, H.A., Johnson, D. and McMurray, W.C., Arch. Biochem. Biophys., 78, 587 (1958).

2. Ernster, L., in Proc. Vth Intern. Congr. Biochem., Moscow, 1961, Pergamon Press, Oxford (in press).

3. Snoswell, A.M., Biochim. Biophys. Acta, 60, 143 (1962).

TOPOGRAPHY OF COUPLING FACTORS IN OXIDATIVE PHOSPHORYLATION

Efraim Racker
The Public Health Research Institute of The
City of New York, Inc.
New York, New York

In planning for this talk I went through a curious process, which reminds me of an old story which is very long and which I won't have time to tell you properly, but I am going to give you a summary. It is the story of a man on a successful business trip, who decides to send a telegram to his wife. First he composes a very long one in which he reports on the success of his business trip, on his social activities and his health, then he inquires about his children and the health of his wife, and finally he remembers that he needs a pair of new shoe laces and he asks his wife to send them to him. Then it dawns on him that this telegram is going to be very expensive, and he proceeds logically, step by step, to eliminate one item after another. The story ends with his sending off this telegram: "Don't send me the shoe laces."

When I started to prepare this talk I went through a similar process. After assembling the experimental material, I argued: the first experiment has already been published in the Journal of Biological Chemistry; the second experiment I reported to Dr. Chance over the telephone and I am sure he told everybody else; the third experiment is one of Tom Conover's and everybody must know it by now; the fourth experiment I am going to present at the Federation Meetings in Atlantic City next week, so why bother showing it twice? Well, in desperation, I went out and bought myself a pair of shoe laces. Finally, I decided that since I would face a very sophisticated audience here, which is keenly interested in the reversal of reactions, I would take my talk for the Federation Meetings and present it in reverse and so complex that I hope nobody will recognize it.

Figure 1, which indicates the topography of coupling factors, I shall not even show at the Federation Symposium. It is strictly for a sophisticated audience.

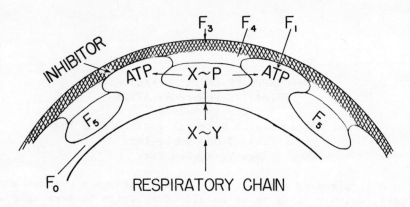

Figure 1. Proposed Topography of Coupling Factors.

The organization of the factors in this figure is based mainly on the characteristics of a variety of derivative particles which differ from each other in the following properties:

1) Dependency on coupling factors for oxidative phosphorylation;
2) Presence of oligomycin-sensitive ATPase;
3) Susceptibility of various factors to the action of trypsin.

The first particles I should like to discuss are the T-U-particles. They are prepared by first exposing mitochondria to sonic oscillation, and by isolating submitochondrial particles (S-particles). These particles are treated with trypsin, then with urea, and are finally re-isolated and suspended in sucrose (1). The resulting T-U-particles contain no ATPase activity, and they do not catalyze oxidative phosphorylation or a P_i^{32}-ATP exchange. However, they contain F_O - the oligomycin-sensitivity factor (2). Soluble ATPase, which is completely resistant to oligomycin, becomes oligomycin-sensitive when added to these particles, as shown in Table I. Trypsin, which has no effect on the oligomycin sensitivity of ATPase in S-particles, rapidly destroys F_O in T-U-particles. Addition of ATPase to T-U-particles confers protection against trypsin. As shown in Table II, it is possible to extract from T-U-particles (but not from S-particles) by sonic oscillation a soluble component which confers oligomycin sensitivity on ATPase. Soluble F_O has some similarity in property to F_4 and to structural protein. But neither F_4 nor structural protein have any effect on the oligomycin-sensitivity of ATPase.

TABLE I
CONFERRAL OF OLIGOMYCIN SENSITIVITY ON SOLUBLE ATPase BY SUBMITOCHONDRIAL PARTICLES TREATED WITH TRYPSIN AND UREA

Additions	ATPase activity μmoles P_i cleaved/10 min		Inhibition
	-oligomycin	+oligomycin	
Exp. 1 T-U-particles (680 μg)	0.16	0.14	
" " + ATPase (3 μg)	1.54	0.3	80 %
Exp. 2a T-U-particles (400 μg) after trypsin treatment + ATPase (3 μg)	1.52	1.40	8 %
2b T-U-particles (400 μg) after trypsin treatment in the presence of ATPase	1.55	0.48	69 %
2c Control for 2a, without trypsin	1.47	0.1	93 %
2d Control for 2b, without trypsin	1.52	0.14	91 %

T-U-particles were prepared by exposing S-particles to trypsin for 30 minutes and then to 2 M urea (cf. Table V). In Experiment 2 the T-U-particles (8mg/ml) were exposed to trypsin for 60 minutes, as described in Table V, except that 120 μg/ml trypsin were used (Experiment 2a). In Experiment 2b, soluble ATPase (60 μg/ml) was added before trypsin. For the control experiments (2c and 2d), the T-U-particles were incubated without and with ATPase, as for 2a and 2b, but without trypsin. ATPase was added at the end of the 60 minutes to Experiments 2a and 2c, and all samples were analyzed for ATPase activity in the absence and presence of oligomycin (3.3 μg/ml).

TABLE II
CONFERRAL OF OLIGOMYCIN SENSITIVITY ON SOLUBLE ATPase BY AN
EXTRACT FROM SUBMITOCHONDRIAL PARTICLES AFTER TREATMENT WITH
TRYPSIN AND UREA

Additions	μmoles P_i cleaved/10 min		Inhibition
	-oligomycin	+oligomycin	
ATPase (3 μg)	1.0	1.06	0
ATPase (3 μg) + F_o (320 μg)	1.18	0.16	86 %
" " + F_3 (1.2 mg)	0.73	0.71	0
" " + F_4 (725 μg)	1.19	1.20	0

The extract containing F_0 was prepared by exposing T-U-particles in 0.25 M sucrose to sonic oscillation for 4 min in a 10 Kc Raytheon oscillator and by centrifugation for 2 hours at 105,000 x g. The supernatant solution was carefully pipetted off and used as a source for F_0. Preparations of F_3 and F_4 were as described previously (3). Oligomycin was added to a final concentration of 3.3 μg/ml.

The second particles I should like to discuss are the P-particles which were studied in collaboration with Dr. Thomas Conover and Dr. Richard Prairie (3, 4). These particles are prepared by sonic oscillation of mitochondria in the presence of 2% phosphatides. They catalyze the oxidation of succinate and DPNH, but do not catalyze phosphorylation, P_i^{32}-ATP exchange, or reversal of electron transport. However, when F_1 and F_4 are added, these catalytic properties are restored to the particles. In Table III, the dependency of phosphorylation on F_4 is shown. It can be seen that phosphorylation due to oxidation of DPN-linked substrates as well as of succinate requires coupling factor 4. More phosphate was esterified with DPN-linked substrates than with succinate, even when the rate of oxidation of the latter was brought to a comparable rate by addition of malonate. This finding indicated that site 1 also requires F_4. This conclusion was confirmed by measurements of phosphorylation in the presence of Antimycin A and phenazine methosulfate, as well as by studies on the reversal of oxidative phosphorylation. P-particles exhibit a similar dependency on F_1 (not shown in Table III). Coupling factor 3 has no effect on oxidative phosphorylation, but gives rise to a 2-fold stimulation of the $P_i{}^{32}$-ATP exchange in P-particles.

TABLE III

DEPENDENCY OF OXIDATIVE PHOSPHORYLATION ON COUPLING FACTOR 4

System	Substrate	Additions	O_2 uptake	P:O
			µatoms/20 min/mg	
Complete	Succinate	-	6.3	0.39
"	"	Malonate (2 mM)	2.4	0.43
F_4 omitted	"	-	6.6	0.05
Complete	Malate + DPN	-	3.6	0.56
"	" + "	Malonate (2 mM)	3.2	0.68
F_4 omitted	" + "	-	3.3	0.03

The complete system contained 1.0 mg of P-particles, 33 µg of F_1, 600 µg of F_4, 8.8 mM $MgCl_2$, and 20 mM succinate, as well as the usual mixture for measurements of oxidative phosphorylation (4). In the experiments with DPN-linked substrates succinate was replaced by malate (40 mM), DPN (0.2 mM), and cysteine sulfinate (40 mM).

In Table IV the dependency of the ATP-dependent reduction of DPN by succinate is shown. A dependency of this reaction, which is catalyzed at site 1 on F_4 and F_1, is clearly demonstrated.

P-particles are therefore visualized in Figure 1 as being stripped off at the level of the F_4 and F_1 layer. This leaves on the surface F_5 (the latest addition to the various coupling factors). This factor stimulates the $P_i{}^{32}$-ATP exchange in P-N-particles (P-particles treated in a Nossal shaker). It is shown in Figure 1 covering F_0, only because in P-particles F_0 is trypsin-resistant, whereas the $P_i{}^{32}$-ATP exchange reaction is highly trypsin-sensitive.

TABLE IV
DEPENDENCY OF DPN REDUCTION BY SUCCINATE ON SOLUBLE PROTEIN FACTORS

Additions	mµmoles DPNH/minute
Complete system	11.2
- submitochondrial particles	0.2
- F_1	0.3
- F_1 + cold-inactivated F_1	0.5
- F_4	0.7
- F_3	5.5
- Bovine serum albumin	4.3
- $MgSO_4$	0.0
- Na_2 succinate	0.0
+ 0.33 µg oligomycin	0.2

The formation of DPNH in P-particles was measured fluorimetrically in a reaction mixture described previously (3).

TABLE V
EFFECT OF TRYPSIN ON ATPase ACTIVITY OF SUBMITOCHONDRIAL PARTICLES

Trypsin treatment minutes	µmoles P_i/10 min/mg protein S-particles	U-particles
0	4.8	0.26
30	43	2.8
60	50	3.9
90	56	4.1

S-particles, prepared as described previously (1), were suspended in 0.05 M Tris buffer, pH 8.0, at 10 mg protein/ml, and exposed at 30° to 30 µg trypsin/ml for various time periods, as indicated. The reaction was stopped by addition of 100 µg trypsin inhibitor/ml. U-particles were prepared by exposure of S-particles to 2 M urea at 0° for 45 min, centrifugation for 30 min at 105,000 x g, washing, and suspending in 0.25 M sucrose. ATPase was measured as described previously (7).

Finally, just a few words about T-particles. These are S-particles treated with trypsin (1). They are visualized in Figure 1 as being stripped off the surface layer. As shown in Table V, a marked stimulation of ATPase activity can be observed, presumably due to inactivation of the ATPase inhibitor (5). Addition of F_3 preparations stimulates the P_i^{32}-ATP exchange and the ATP-dependent reduction of DPN by succinate in these particles.

Where do we go from here? We would like to see a complete reconstruction of the phosphorylating system with soluble factors, but thus far we have not been successful. Coupling factors F_4 and F_0, both of which have some similarity to the structural protein of Criddle and his collaborators (6), may be involved in the structural organization of the phosphorylating system. I should like to point out that the structural protein, which has been shown to interact with members of the respiratory chain, does not confer oligomycin sensitivity on ATPase, and does not substitute for F_4. But we may find eventually that these different structural proteins are sisters under the skin and fit together in the organizational jig-saw puzzle. What we hope to learn from our experiments, in which we strip off the various layers of the submitochondria, is some insight into the organizational pattern of the interacting factors. I hope one day we will be able to say: Even this strip-tease is a reversible reaction.

REFERENCES

1. Racker, E., Proc. Natl. Acad. Sci. U. S. 48, 1659 (1962).
2. Racker, E., Biochem. Biophys. Res. Communs. 10, 435 (1963).
3. Prairie, R. L., Conover, T. E., and Racker, E., Biochem. Biophys. Res. Communs. 10, 422 (1963).
4. Conover, T. E., Prairie, R. L., and Racker, E., J. Biol. Chem. (in press).
5. Pullman, M. E., and Garber, E. R., Proc. Vth Int. Congr. Biochem., Moscow 1961. London, Pergamon Press, Ltd., 1963, p. 470.
6. Criddle, R. A., Bock, R. M., Green, D. E., and Tisdale, H., Biochem. 1, 827 (1962).
7. Pullman, M. E., Penefsky, H. S., Datta, A., and Racker, E., J. Biol. Chem. 235, 3322 (1960).

DISCUSSION

Estabrook: Are you suggesting that such a scheme is appropriate for each site of phosphorylation, or merely that there are some points which are in common?

Racker: Our present feeling is that this only applies to Site 1 and Site 2.

Chance: I think Dr. Estabrook wants to know where the elementary particles are in your diagram, Dr. Racker.

Racker: We have carefully omitted them.

King: We have done some work very similar to Selwyn and Chappell's on the F_0 and F_1 in the old-fashioned Keilin-Hartree preparation. We have found that the ATP-ase in the Keilin-Hartree preparations can be solubilized, and can be put back to form an oligomycin-sensitive preparation. However I would like to ask whether urea dissociates F_1, or just inactivates it?

Racker: We think it probably inactivates and dissociates the protein.

King: First dissociates or first inactivates?

Racker: It inactivates it much faster. The soluble ATP-ase is very sensitive to low concentrations of urea in the cold, but not so much at room temperature. As to the rate of inactivation: in the absence of urea, the half-life at $0°$ is about two hours, whereas in the presence of 0.8 urea, the soluble ATP-ase has a half-life of about seven minutes.

King: Does that "dead" ATP-ase remain in the particles?

Racker: We don't think so; it is not there and we cannot test for it.

King: Functionally or structurally?

Racker: We think it comes off, because we can restore activity by addition of active F_1 (ATP-ase). Incidentally, one thing which will interest you, Dr. King, is that the Keilin-Hartree preparation, which we have prepared according to your procedure, but in the cold instead of at room temperature, catalyzes a P_i-ATP exchange reaction which is oligomycin-sensitive. Therefore, they still have some phosphorylating capaci

King: Therefore the term "non-phosphorylating particles" is now less meaningful.

Racker: Yes.

Brierley: Structural protein (S.P.) has been prepared by several different procedures in our laboratory. Although the bulk of the protein is inert, some of these preparations exhibit detectable enzymatic activity, such as B-hydroxybutyric dehydrogenase and α-ketoglutaric dehydrogenase (1). These activities are lost following removal of the lipid and deoxycholate by the published procedure (2). Less drastic procedures are now available for the preparation of S.P. (3) and it would be of interest to compare S.P. prepared by this method with Dr. Racker's preparation.

Sanadi: The three factors you have shown in the succinate-linked DPN reduction are F_1 and F_4 absolutely, and F_3 partially—is that correct?

Racker: Yes.

Sanadi: Could the requirement for F_3 be replaced by bovine serum albumin?

Racker: No, our data show that the effects of F_3 and serum albumin are independent; in fact, all of our studies on the ATP-dependent reduction of DPN have been carried out in the presence of bovine serum albumin.

Chappell: Where do you visualize the site of action of dinitrophenol in your system?

Racker: Mildred Cohn and her collaborators have shown that dinitrophenol acts as if at low concentrations it hydrolyzes an $X \sim Y$ intermediate and at high concentrations an $X \sim P$ intermediate. There are three possibilities: we can propose that dinitrophenol catalyzes the hydrolysis of $X \sim Y$, or of $X \sim Y$ and $X \sim P$, or that it catalyzes neither. The last possibility one can formulate as follows: dinitrophenol interacts with coupling factor 1 and changes its structural conformation in such a manner that it no longer attaches properly to the respiratory chain. Now water can enter to hydrolyze on the one hand, $X \sim P$ (ATP-ase) and on the other hand, $X \sim Y$ (loss of respiratory control). When coupling factor and respiratory chain are tightly coupled, water cannot enter into either reaction. According to this formulation, you see dinitrophenol has nothing to do with hydrolysis itself, but only affects the interaction between F_1 and the respiratory chain.

Webster: Which of these particles, TU and such, can you

restore phosphate esterification? Does it take all five of the factors in TU particles?

Racker: I have so far not been able to restore the TU particles, but I am still trying. We can completely restore the P-particles.

Webster: With the combination?

Racker: With the combination of F_1 and F_4. Only these two are really required for net phosphorylation; F_3, however, stimulates the $P_i{}^{32}$-ATP exchange.

King: In your latest BBRC paper, it seems to me that ATP-ase activity is lower than you previously reported. But in Table V of your paper, it is 4.8 units for 10 minutes.

Racker: That is what we find in particles obtained by sonic oscillation of beef heart mitochondria.

King: That is only 0.48 per minute. Is that a regenerating system?

Racker: This is the rate we usually observe, but ATP-ase activity varies very much, depending on how you prepare the particles.

King: Is ATP-ase activity increased when you add F_1 to the normal particles, instead of urea-treated particles?

Racker: When soluble ATP-ase is added to N-particles (obtain in the Nossal shaker) the two activities do not add up. Dr. Pullman has observed as much as 90 per cent masking of the soluble ATP-ase activity after addition to particles. The activity of the soluble ATP-ase cannot be recovered after the coenzyme combines with the particles.

Pullman: You have an inhibition of the added ATP-ase.

Estabrook: You have given us six factors so far, and you have not mentioned, except briefly, the coupling factor--the respiratory control factor. Is this another one which must be included in the scheme?

Racker: It is possible that the inhibitor of ATP-ase which Dr. Pullman has studied extensively during the past year has something to do with respiratory control. This inhibitor is a rather interesting protein. It is heat stable, it is precipitated by trichloroacetic acid and redissolved without losing activity, it is soluble in 80 per cent ethanol, but is highly sensitive to proteolytic digestion. When added to submitochondrial particles it eliminates ATP-ase activity without interfering with oxidative phosphorylation. However, particles so treated do not have respiratory control. Our present view therefore, is that the ATP-ase inhibitor may be part of the mechanism of respiratory control, but is not the whole story.

REFERENCES

1. Blair, P. V., unpublished, and Jurtshuk, unpublished.
2. Criddle, W. J., et al., Biochem., $\underline{1}$, 827 (1962).
3. Richardson, S. H., Hultin, H., and Fleischer, S., in preparation.

THE ENERGY-LINKED REDUCTION OF UBIQUINONE IN BEEF HEART MITOCHONDRIA
Harvey S. Penefsky

Considerable evidence has accumulated which supports the concept of energy-linked reversal of electron transport from reduced cytochrome c to pyridine nucleotide first described by Chance and Hollunger (1) and by Chance and Fugmann (2). Further study of this reaction has been greatly facilitated by the use of the dye tetramethyl-p-phenylenediamine (TMPD) as an external source of electrons. As shown by Jacobs (3), the oxidation of ascorbate, mediated by catalytic concentrations of TMPD, is efficiently coupled to the synthesis of ATP from ADP and inorganic phosphate. The dye was thought to enter the respiratory chain by reducing endogenous cytochrome c. The reduction of DPN^+ associated with the oxidation of TMPD was shown by Packer (4) and has been confirmed by other workers with a variety of mitochondrial and submitochondrial preparations (5, 6).

The reduction of ubiquinone during the oxidation of TMPD by beef-heart mitochondria has been reported (7). Heavy layer mitochondria used in these experiments were prepared as described by Hatefi and Lester (8). The reactions were terminated with cold methanol and the ubiquinone extracted into isooctane by a slight modification of the method of Pumphrey and Redfearn (9, cf. ref. 7). After evaporation of the isooctane extracts, ubiquinone was dissolved in ethanol and determined by the change in absorbancy at 275 mμ before and after the addition of potassium borohydride (10). Since the borohydride method estimates only the amount of oxidized ubiquinone present, each experiment was carried out in duplicate. Total ubiquinone (oxidized plus reduced) was estimated in the second sample after oxidation of ubiquinol by $AuCl_3$ (11). Ubiquinone reduced in each experiment was then calculated as the differ-

---

Present address: Public Health Research Institute of the City of New York, New York, New York.

This investigation has been aided in part by grants from the Jane Coffin Childs Memorial Fund for Medical Research and from the Life Insurance Medical Research Fund.

ence between the two samples. The $AuCl_3$ oxidation step served as a check on the efficiency of extraction of ubiquinone under varying experimental conditions.

TABLE I
THE REDUCTION OF UBIQUINONE OF BEEF-HEART MITOCHONDRIA BY TMPD

Additions	A Total ubiquinone	B Ubiquinone (mµmoles)	Ubiquinol*	Reduction (%)
None (substrate omitted)	21.7	22.1	--	0
None (zero-time control)	22.0	22.5	--	0
None	23.2	12.7	10.5	45.2
Antimycin A (13.4 µg)**	23.4	21.9	1.5	6.4
2,4-dinitrophenol (0.2 mM)	23.7	22.6	1.1	4.6
Oligomycin (10 µg)	23.6	12.8	10.8	45.7

* Ubiquinol = total ubiquinone minus ubiquinone.

** Antimycin A was added after the 2 min incubation at 30°. The incubation was continued for an additional 60 sec with antimycin before cooling the reaction mixture to 0°.

The reaction mixture, in a final volume of 1.0 ml, contained 40 mM Tris-HCl (pH 7.4), 2.45 mM sodium amytal, 0.3 mM TMPD, 10 mM ascorbate and 150-210 mM sucrose. The reaction was initiated by the addition of 0.15 ml of a solution containing TMPD and ascorbate. Oligomycin and antimycin were added in 0.2 ml ethanol. 5.9 mg mitochondria were incubated with the additions shown for two min at 30° with vigorous shaking to oxidize any endogenous ubiquinol. The reaction mixture was then cooled to 0° by immersion for 30 sec in an ice bath. Substrate was added and the incubation was continued for 20 sec at 0°.

Table I summarizes the evidence which led us to believe that reduction of endogenous ubiquinone during the oxidation of TMPD occurs via an energy-linked reversal of the respiratory chain. Ubiquinone was not reduced in the absence of added substrate or in the zero time control. Endogenous ubiquinone was almost entirely oxidized before the addition of substrate. In the presence of TMPD, ubiquinone was about 45% reduced. The reduction was inhibited by dinitrophenol and antimycin but not by oligomycin. In separate experiments, arsenate was shown to cause partial inhibition of the reaction. Since amytal also was present, endogenous substrate probably did not contribute reducing equivalents to ubiquinone. Thus the sensitivity of TMPD-linked reduction of ubiquinone to uncouplers of oxidative phosphorylation and to antimycin and the

(text continued p. 90)

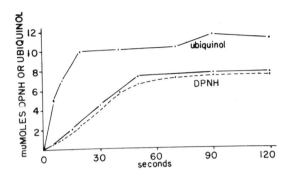

Fig.1. Comparison of the rates of reduction of ubiquinone and of DPN$^+$ in beef-heart mitochondria by TMPD. The complete reaction mixture contained, in a volume of 2.0 ml, 40 mM Tris-HCl (pH 7.4), 210 mM sucrose, 5.4 mg mitochondria, 0.3 mM TMPD and 10 mM ascorbate. The sarcosomes were incubated at 30° for 2 min with vigorous shaking to oxidize any DPNH or ubiquinol and then cooled to 0° by immersion for 1 min in an ice bath. The reaction was initiated by the addition of 0.15 ml of a solution containing TMPD and ascorbate and was allowed to continue for varying periods of time at 0°. When ubiquinone was to be estimated, the reaction was stopped with cold methanol and ubiquinone was determined as described in the text. AuCl$_3$ oxidation (to determine the total ubiquinone of the sample) was performed only on the zero time control. The amounts of ubiquinol shown in the curve were calculated as the difference between the total ubiquinone in the zero time control and the amount of ubiquinone found at each point of the incubation. Measurements of DPN$^+$ reduction, carried out separately, were made in two ways. DPN$^+$ and DPNH were estimated in acid and alkaline extracts, respectively, of the mitochondria according to the method of Purvis[13]. DPNH was determined only in the zero time control and after 120 sec of incubation with substrate. DPN$^+$ was determined for all points of the curve. In the latter case, the reaction was stopped by rapidly injecting into the reaction mixture 0.2 ml of 50% trichloroacetic acid. DPNH shown at each point of the solid curve was calculated as the difference between the amount of DPN$^+$ found in the sample and the total DPN$^+$ of the zero time control. Calculated DPNH at 120 sec and DPNH found at this point were in close agreement. DPN$^+$ reduction in the mitochondria also was monitored continuously in an Eppendorf fluorometer[14]. The dashed curve is a fluorometer tracing which was calibrated with reference to mitochondrial DPNH values as determined by the method of Purvis[13].

lack of sensitivity to amytal and oligomycin are observations which support the participation of ubiquinone in an energy-linked reversal of electron transfer from cytochrome c in the respiratory chain.

Chance has shown (12) that in phosphorylating pigeon-heart mitochondria, the reduction of DPN^+ by succinate occurs at about the same rate as the reduction of ubiquinone. It was of interest, therefore, to compare the rates of TMPD-linked reduction of ubiquinone and of DPN^+ under identical conditions. Such a comparison is shown in Fig. 1. It may be seen that ubiquinone reduction was considerably faster than DPN^+ reduction under these conditions, and in general was three to five times faster. These results are consistent with the assignment of a position of ubiquinone in the main path of electron transfer between cytochrome c and DPN^+. However, these results do not rule out the possibility that ubiquinone occupies a position on a side path which is in rapid equilibrium with a component of the main pathway.

REFERENCES

1. Chance, B., and Hollunger, G., J. Biol. Chem., 236, 1534 (1961).

2. Chance, B., and Fugmann, U., Biochem. Biophys. Res. Comm., 4, 317 (1961).

3. Jacobs, E. E., Biochem. Biophys. Res. Comm., 3, 536 (1960).

4. Packer, L., J. Biol. Chem., 237, 1327 (1962).

5. Tager, J. M., Howland, J., and Slater, E. C., Biochim. Biophys. Acta, 58, 616 (1962).

6. Löw, H., and Vallin, I., Biochem. Biophys. Res. Comm., 9, 307 (1962).

7. Penefsky, H. S., Biochim. Biophys. Acta, 58, 619 (1962).

8. Hatefi, Y., and Lester, R. L., Biochim. Biophys. Acta, 27, 83 (1958).

9. Pumphrey, A. M., and Redfearn, E. R., Biochem. J., 76, 61 (1960).

10. Crane, F. L., Lester, R. L., Widmer, C., and Hatefi, Y., Biochim. Biophys. Acta, 32, 73 (1959).

11. Bouman, J., Slater, E. C., Rudney, H., and Links, J., Biochim. Biophys. Acta, 29, 456 (1958).

12. Chance, B., in *A Ciba Foundation Symposium on Quinones in Electron Transport*, 1960, Churchill Ltd., London, 1961, p. 327.

13. Purvis, J., Biochim. Biophys. Acta, 38, 435 (1960).
14. Estabrook, R. W., Anal. Biochem., 4, 231 (1962).

DISCUSSION

<u>Packer</u>: What per cent of the ubiquinone and pyridine nucleotide is reducible by the system? Can you assume that the pools all react as one in both cases?

<u>Penefsky</u>: The total ubiquinone reducible by this system is about 75 per cent. This compares with the total of about 80-85 per cent reducible in the presence of succinate and Antimycin. The total DPN reducible under these conditions is about 60 per cent.

<u>Klingenberg</u>: The reduction of ubiquinone appears to be energy-dependent in a different way than just by reversed electron transport. After uncoupling of the phosphorylation with calcium, ubiquinone is slowly reduced on addition of Antimycin:

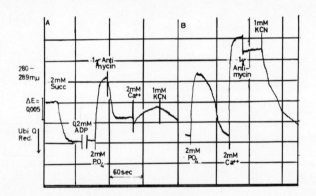

Fig. 1. The redox state of ubiquinone in mitochondria in response to Antimycin. Recording of ubiquinone absorption in a suspension of heart mitochondria. A. On addition of Antimycin to the active state, ubiquinone is rapidly reduced. B. On addition of Antimycin to mitochondria "uncoupled" by Ca^{++}, no reduction is recorded. The downward deflection on anaerobiosis is due to the absorption of Antimycin.

Here the effect of Antimycin addition to phosphorylating heart muscle mitochondria is shown. Ubiquinone reduction is a downward deflection. The reduction with succinate is rather slow.

Then a cyclic oxidation follows the addition of ATP and phosphate, which is further followed by a fast reduction on addition of Antimycin. On repeating the experiment, ubiquinone oxidized by addition of calcium was not reduced by Antimycin.

In this respect ubiquinone is similar to DPN or to a large part of the flavoprotein which can no longer be reduced on addition of Antimycin (1), or sometimes even in anaerobiosis when phosphorylation has been uncoupled.

Penefsky: Is endogenous substrate reducing ubiquinone?

Slater: This is succinate.

Chance: Succinate reducing ubiquinone?

Klingenberg: Yes, succinate reduces the ubiquinone in mitochondria.

Packer: I should think that 2 mM calcium would tear the mitochondria apart.

Chance: Certainly we agree with Penefsky that the kinetics and inhibitor sensitivity of the endogenous ubiquinone reduction is very similar to that of DPN.

Racker: How about the rate?

Chance: The rate is equally slow. In pigeon heart mitochondria, at least, we do not find the much faster ubiquinone reduction that you got in your system. It is a very slow reaction (2).

Hess: Is the rate of oxidation in the presence of calcium phosphate-dependent?

Klingenberg: Yes.

Griffiths: Would the analytical method tell you whether any ubichromenol or ubichromenol phosphate is formed in that reaction?

Penefsky: We were very much interested in this, but the analytical methods were such that we could not tell. I might add that I have looked for the formation of ubichromenol under a variety of conditions in beef-heart sarcosomes but was not able to detect it. I understand that Dr. Karl Folkers also was not able to find ubichromenol on beef-heart sarcosomes.

Roy: Is there any non-enzymatic reduction of ubiquinone, and how did you take care of that? I think that in Dr. Sanadi's system, ascorbate reduced quinones non-enzymatically.

Penefsky: I was not able to reduce aqueous suspensions of crystalline ubiquinone (50) by ascorbate or by ascorbate and TMPD.

REFERENCES

1. Klingenberg, M., and Bucher, Th., Biochem. Z., <u>331</u>, 332 (1959).
2. Chance, B., and Hagihara, B., in Proc. Fifth Intern. Cong. Biochem., Moscow, 1961, Pergammon Press, Oxford, in press.

THE TRANSFER OF REDUCING POWER

PROVISION OF REDUCING POWER FOR GLUTAMATE SYNTHESIS

E.C. Slater and J.M. Tager

Laboratory of Physiological Chemistry, University of Amsterdam
The Netherlands

It has been known for about 30 years that the energy made available by exergonic oxidation reactions can be trapped for the formation of high-energy compounds.

$$AH_2 + B \longrightarrow A + BH_2 + \text{energy}$$

$$AH_2 + B \rightleftharpoons A + BH_2 +$$

Fig. 1

trapping of this energy would be expected to make the oxidation-reduction reaction more easily reversible. This was easily demonstrable with the glyceraldehyde phosphate dehydrogenase reaction (1), but its demonstration for respiratory chain phosphorylation proved difficult until Dr. Chance showed us how to do it.

He and Hollunger showed some time ago that succinate added to mitochondria in the absence of ADP brought about the rapid and extensive reduction of the mitochondrial pyridine nucleotide (2, 3). Further studies by Chance and Hollunger (4-6), by Klingenberg (7, 8) who introduced the related but in many respects less complicated substrate glycerol phosphate into these studies, by Ernster (9, 10) and by Snoswell (11, 12), Penefsky (13) and others (14-16) in our own group, have firmly established the chief characteristics of this reaction. Those assembled in this colloquium hardly need convincing that the reaction involves an energy-linked reversal of the respiratory chain.

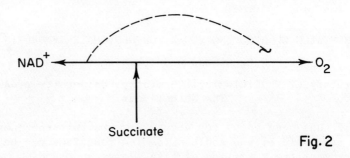

Fig. 2

Fig. 2 shows schematically the essential features of the mechanism. The oxidation of succinate by oxygen is coupled with the oxidation of succinate by NAD^+. Or in other words, the energy generated in the oxidation of succinate by O_2 can be utilized, not only for the synthesis of ATP from P_i and ADP, but also for the energy-requiring reduction of NAD^+ by succinate.

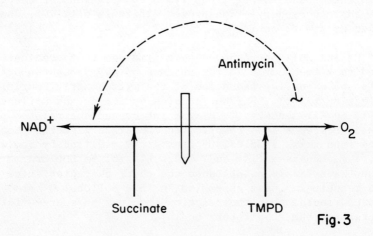

Fig. 3

Here succinate provides reducing equivalents both to NAD^+ and to oxygen. This is not, however, necessary. We can replace succinate as donor for the energy-generating reaction as shown in Fig. 3. Antimycin blocks the energy-generating reaction with succinate as donor, without affecting its ability to reduce NAD^+ when energy is supplied to the system. The aerobic oxidation of tetramethyl-p-phenylenediamine, which enters the respiratory chain at about the level of cytochrome c, can supply this energy (15-18).

TABLE I

SYNTHESIS OF GLUTAMATE IN SUCCINATE-ANTIMYCIN-TMPD SYSTEM

KCl, EDTA, Tris, P_i, $MgCl_2$, NH_4Cl, α-ketoglutarate, As_2O_3, ADP, glucose, hexokinase, oligomycin present in all cases
[TAGER, HOWLAND AND SLATER]

EXPT.	ADDITION	ΔO(μatoms)	ΔGlu(μmoles)
1	NONE	0.18	0
	ANTIMYCIN	0.18	0
	SUCCINATE	4.63	4.27
	SUCCINATE + ANTIMYCIN	0.13	0.11
	SUCCINATE + ANTIMYCIN + TMPD	7.2	1.87
2	SUCCINATE + ANTIMYCIN + TMPD	12.0	1.74
	SUCCINATE + ANTIMYCIN + TMPD + OLIGOMYCIN	11.3	2.81

This is illustrated in Table I (cf. ref. 15). The reduction of NAD^+ was followed by measuring the reduction of α-oxoglutarate (+ NH_3) to glutamate.

TMPD can also provide reducing equivalents. The results of Snoswell (18) shown in Table II illustrate the reduction

TABLE II

REDUCTION OF NAD^+ IN RABBIT-HEART SARCOSOMES BY SUCCINATE AND TMPD

	Snoswell		90 sec, 0°	
	NAD^+	NADH	NAD^+ + NADH	$\frac{NADH}{NAD^+ + NADH}$
Fresh sarcosomes	4.5	2.9	7.4	0.39
Incubated with TMPD	0.7	6.4	7.1	0.91
Incubated with succinate	1.2	5.1	6.3	0.80

of NAD^+ by TMPD coupled with the oxidation of TMPD by oxygen (cf. ref. 19-22), and those of Penefsky (13) shown in Table III, the reduction of ubiquinone by TMPD coupled with the

TABLE III

REDUCTION OF UBIQUINONE COUPLED WITH OXIDATION OF TMPD
Reaction at 0°   [PENEFSKY]

ADDITION	REACTION TIME (sec)	UQ (mµmoles)	UQH$_2$ (mµmoles)	% REDUCTION
NONE	0	22.5	-	0
NONE	20	22.1	-	0
TMPD	20	12.7	10.5	45
TMPD + ANTIMYCIN	20	21.9	1.5	6
TMPD + OLIGOMYCIN	20	12.8	10.8	45

oxidation of TMPD by oxygen. This reaction is very rapid. We can describe these reactions by the scheme shown in Fig. 4.

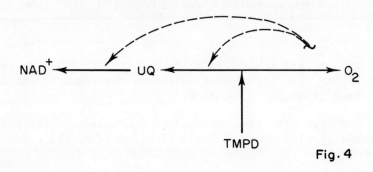

Fig. 4

One of the most interesting findings in this field was made independently and at about the same time by Ernster (9, 10) in Stockholm and by Snoswell (11, 12) in Amsterdam, viz. that oligomycin, which inhibits the synthesis of ATP linked to the oxidation of succinate or TMPD by oxygen, has no effect on the utilization of this energy for reversing the respiratory chain. This was the first experimental demonstration in favor of a proposal made 10 years ago (23) that the energy of intermediates of oxidative phosphorylation might be utilized directly for energy-requiring reactions without conversion to ATP being necessary. Ernster (10) provided further important confirmation of this concept when he demonstrated that the reduction of acetoacetate by succinate (presumably via NAD$^+$)

coupled with the oxidation of succinate by oxygen could proceed in the absence of phosphate. This was confirmed by Snoswell (12) who found that the reduction of mitochondrial NAD^+ linked with the aerobic oxidation of succinate proceeded readily in the absence of inorganic phosphate.

All the reactions which we have described to date can be carried out - indeed are better carried out - in the presence of oligomycin, the high-energy compounds being generated by the operation of the respiratory chain - internally generated, as Chance calls it. Chance (5) and Klingenberg (24) have demonstrated that ATP added from outside can also provide the energy when the cytochrome oxidase is inhibited. Low (25, 26) has utilized this reaction to demonstrate the reduction of externally added NAD^+, using sub-mitochondrial particles as the catalyst.

ATP can also be used in our system. Table IV illustrates

TABLE IV

ATP-STIMULATED GLUTAMATE SYNTHESIS

SUCCINATE, α-KETOGLUTARATE, NH_4Cl, As_2O_3, low P_i (2mM)

FURTHER ADDITIONS	ΔO (µatoms)	ΔGlu (µmoles)
NONE	7.2	2.3
OLIGOMYCIN	8.6	5.6
ANTIMYCIN	0.7	0.7
ANTIMYCIN + OLIGOMYCIN	0.7	0.5
ANTIMYCIN + ATP	0.7	1.6
ANTIMYCIN + ATP + OLIGOMYCIN	0.6	0.5

the ATP requirement for NAD^+ reduction (again measured by glutamate synthesis) and the sensitivity to oligomycin. This is shown schematically in Fig. 5. The oligomycin sensitivity of the effect of ATP is diagnostic that ATP is reacting by this type of mechanism.

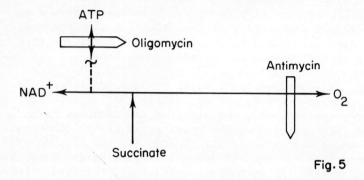

Fig. 5

None of the points discussed to date are controversial, although perhaps others might put the emphasis somewhat differently. We now wish to fulfill the wish of our chairman by embarking on to less firmly established and therefore more controversial territory.

First, we should like to discuss the value of $\underline{n}$ in Eqn.

$$\text{Succinate} + \text{NAD}^+ + n\sim \longrightarrow \text{Fumarate} + \text{NADH} + \text{H}^+ \qquad (1)$$

This is important for the consideration of the mechanism of the reaction. Two sets of values have been reported in the literature. Chance (5) found a yield of 0.6 molecule NADH/$\sim$, while Ernster (10) found a value of about 1. Since the method of Chance would tend to under-estimate the yield while that of Ernster was based on assumptions which, if not correct, would over-estimate it, we have reinvestigated the matter, using a different method. We found it convenient to oxidize the NADH formed by α-ketoglutarate (+ NH_3) and to measure the glutamate formed. 1 mM arsenite was added to prevent the oxidation of α-ketoglutarate. One must also expect the NAD^+-catalysed reduction of α-ketoglutarate (+ NH_3) by the malate formed from fumarate, the product of oxidation of succinate, especially when glutamate is present to remove the oxaloacetate formed (Fig. 6). Under these conditions, the amount of aspartate

$$\text{MALATE} + \text{NAD}^+ \rightleftharpoons \text{OXALOACETATE} + \text{NADH} + \text{H}^+$$

$$\text{NADH} + \text{α-KETOGLUTARATE} + NH_3 + H^+ \rightleftharpoons \text{NAD}^+ + \text{GLUTAMATE}$$

$$\text{GLUTAMATE} + \text{OXALOACETATE} \rightleftharpoons \text{α-KETOGLUTARATE} + \text{ASPARTATE}$$

$$\underline{\text{Sum}} \quad \text{MALATE} + NH_3 \rightleftharpoons \text{ASPARTATE}$$

Fig. 6

formed during the course of an experiment, then, is a measure of the rate of reduction of the mitochondrial NAD^+ by the malate formed from the succinate. More accurately, this is given by (aspartate + oxaloacetate) since oxaloacetate is present at the end of the experiment in amounts equal to rather less than 10% of the amounts of aspartate found. Glutamate does not appear in the sum reaction, so that any glutamate which is present at the end of the reaction (or more accurately glutamate <u>minus</u> oxaloacetate) must have been derived from some other donor. The only other donor available under the conditions of these experiments is succinate.

The (aspartate + oxaloacetate) found in our experiments is a measure of the NAD reduced by the Krebs mechanism (27), the (glutamate-oxaloacetate) by the Chance reaction, which is the one we are now interested in.

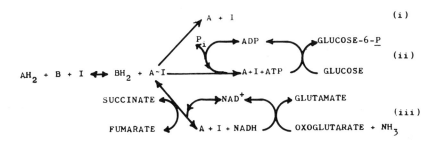

Fig. 7

According to the mechanism of oxidative phosphorylation which we prefer, high-energy compounds (indicated here by $A{\sim}I$) are formed during the operation of the respiratory chain, indicated in Fig. 7 by the oxidation of AH_2 by B, and are decomposed by P_i and ADP in a reaction leading to the synthesis of ATP. In the absence of P_i or ADP, or when the reaction with P_i and ADP is inhibited by oligomycin, respiration is limited by the rate of the spontaneous decomposition of the high-energy compounds. This decomposition can be accelerated, with a corresponding stimulation of the respiration, by the addition of dinitrophenol.

In the presence of oligomycin, respiration is also stimulated by the addition of α-ketoglutarate (+ NH_3), which, by oxidizing the NADH, promotes the energy-requiring reduction of NAD^+ by succinate. This is shown for succinate as substrate

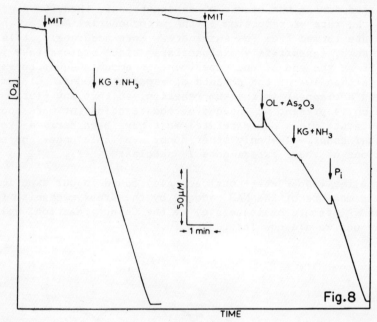

Fig. 8

in Fig. 8. When oligomycin is absent, and ADP, P_i, α-ketoglutarate and NH_3 are all present, there will be the three competing reactions for the ∿ compound -

Pathway i, stimulated by dinitrophenol.

Pathway ii, requiring P_i and ADP and inhibited by oligomycin.

Pathway iii, requiring α-ketoglutarate, NH_3 and succinate and inhibited by Amytal or hexylguanidine.

Since the decomposition by Pathway ii is sufficiently rapid for the oxidation reaction to become rate-limiting, the addition of α-ketoglutarate + NH_3 to a system containing ADP and P_i does not stimulate the rate of oxidation, but by introducing Pathway iii the amount of A∿I available for the synthesis of ATP is decreased. The decrease in ATP synthesis (measured by the amount of glucose-6-P formed) brought about by the addition of α-ketoglutarate + NH_3 gives, then, the amount of A∿I used for the reduction of NAD^+ leading to the synthesis of glutamate. The yield of glutamate can be expressed by the ratio Δ glutamate:-Δ∿, where -Δ∿ = (glucose-6-P in the presence of α-ketoglutarate + NH_3) minus (glucose-6-P in the absence of α-ketoglutarate + NH_3).

TABLE V

DETERMINATION OF $\Delta GLU:-\Delta\sim$

NH_4Cl, succinate, arsenite, glucose, hexokinase, ADP, EDTA, $MgCl_2$

	Expt. 173		Expt. 195			
α-kg (mM)	0	20	0	20	0	20
dinitrophenol	0	0	0	0	6	6
ΔO (μatoms)	9.0	8.7	12.1	13.1	13.1	13.4
Δest.P (μmoles)	12.2	8.7	14.1	10.7	12.3	9.7
P:O	1.36	1.00	1.08	0.82	0.94	0.73
Δest.P (corr*)	11.8	8.7	15.2	10.7	12.6	9.7
Decrease in Δest.P ($-\Delta\sim$)		3.1		4.5		2.9
ΔGlu (μmoles)	0	2.47	0.07	3.60	0.30	3.04
ΔAsp (μmoles)	0.28	1.89	0	3.49	0	2.91
$\Delta OxAc$ (μmole)	0	0	0.02	0.32	0.02	0.31
$\Delta Glu:-\Delta\sim$		0.80		0.78		0.95
Δ(Glu-OxAc): $-\Delta\sim$				0.72		0.85

* For difference between ΔO with and without α-ketoglutarate

Table V shows some typical experiments, yielding Δglutamate:$\sim$ratios of about 0.8, or a little less when the amounts of oxaloacetate found are subtracted.

TABLE VI

$\Delta GLU:-\Delta\sim$ WITH SUCCINATE AS DONOR

Energy Source	$\Delta GLU:-\Delta\sim$		
	No.	Range	Mean
Succinate $\to O_2$	9	0.73 - 1.11	0.84
TMPD $\to O_2$	3	0.49 - 1.06	0.82
ATP	4	0.25 - 0.68	0.48

Table VI shows the mean values obtained with succinate, and also with TMPD. The ratio approaches 1, and as we shall see later it should be corrected upwards towards 1.

We now come to the most controversial question, the mechanism of Reaction 1. Our working mechanism is given in Fig. 9, which is the sort of mechanism which those of you familiar with our work might have expected. I_1, I_2 and I_3 represent components required at the three phosphorylation steps of the respiratory chain. They may or may not be identical.

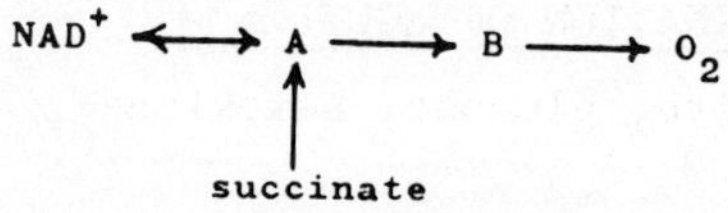

$$NAD^+ \longleftrightarrow A \longrightarrow B \longrightarrow O_2$$
$$\uparrow$$
$$\text{succinate}$$

$$3 \text{ succinate} + 3 A \rightleftharpoons 3 \text{ fumarate} + 3 AH_2$$
$$AH_2 + B + I_2 \rightleftharpoons A\sim I_2 + BH_2$$
$$BH_2 + 1/2\, O_2 + I_3 \longrightarrow B\sim I_3 + H_2O$$
$$A\sim I_2 + NAD^+ + I_1 \rightleftharpoons A + I_2 + NAD\sim I_1$$
$$B\sim I_3 + NAD^+ + I_1 \rightleftharpoons B + I_3 + NAD\sim I_1$$
$$2\, AH_2 + 2\, NAD\sim I_1 \rightleftharpoons 2\, A + 2\, NADH + 2\, I_1 + 2\, H^+$$

Sum $\quad 3 \text{ succinate} + 1/2\, O_2 + 2\, NAD^+ \longrightarrow 3 \text{ fumarate} + 2\, NADH + 2\, H^+$

Fig. 9

A could conceivably be ubiquinone. If this is the case, the first reaction would proceed through the succinate dehydrogenase flavoprotein, and the last through the NADH dehydrogenase. Alternatively, A could be succinate dehydrogenase, and the last reaction involve ubiquinone and NADH dehydrogenase. For either case, it is presumably reduced NADH dehydrogenase flavoprotein which provides the reducing equivalents to the NAD^+. According to our mechanism it is not NAD^+ which is reduced but $NAD\sim I$. Indeed, the first 5 reactions of this mechanism are the same as we proposed about 5 years ago, before we even believed that succinate reduced NAD^+. In 1958, Purvis (28, 29) in our laboratory confirmed Chance and Hollunger's finding that NAD^+ disappeared when succinate was added to rat-liver mitochondria, but he did not find any of the disappearing NAD^+ back as NADH. It just disappeared. In subsequent work from our laboratory (30, 12), much of the disappearing NAD^+ was found as NADH, but a substantial amount remained unaccounted for. If our mechanism is right, this unaccounted-for NAD - the extra NAD of Purvis - would be $NAD\sim I_1$, and the difference between the results obtained by Purvis and his successors (30, 12) could be ascribed to different ratios of $NAD\sim I_1$ and NADH, set up by the last equilibrium.

This mechanism explains the inhibition by dinitrophenol or arsenate, which discharges the high-energy intermediates, and by Amytal and antimycin, which according to Hulsmann's theory (31) combine with $NAD\sim I_1$, and $A\sim I_2$, respectively.

The possible involvement of a high-energy form of NAD^+ in these reactions has recently been supported by work in other directions.

You will recall that in our system with succinate as substrate, malate-reduced NAD turns up as aspartate. It was found that dinitrophenol inhibited not only the synthesis of glutamate, which was to be expected for an energy-linked reaction but also the synthesis of aspartate. This was a surprising result, since it suggested that energy was necessary for the text-book NAD-catalysed oxido-reduction between malate and α-ketoglutarate + NH_3. By using malate as substrate, this could be demonstrated directly (Table VII-cf. ref. 32). In

TABLE VII

MALATE AS HYDROGEN DONOR: EFFECT OF INHIBITORS AND P_i ACCEPTOR

Reaction mixture contained α-oxoglutarate, NH_4Cl, arsenite, malate, glutamate and P_i.

Additions	ΔGlutamate (μmoles)	ΔAspartate (μmoles)	Δ(Glutamate + aspartate) (μmoles)
None	-0.3	7.3	7.0
Oligomycin	-0.1	9.2	9.1
Antimycin	0.9	1.7	2.6
Amytal	0.5	0.9	1.4
Glucose-hexokinase	-2.4	6.9	4.5
Glucose-hexokinase + oligomycin	0.2	7.7	7.9

other experiments, it was shown that the energy could be provided by the aerobic oxidation of malate, of succinate in the presence of Amytal, or of TMPD in the presence of Amytal or antimycin, or by ATP. Klingenberg (33) obtained very similar results at about the same time.

TABLE VIII

DETERMINATION OF Δ(GLU + ASP):-Δ~ WITH MALATE
α-ketoglutarate, malate, arsenite, glucose, hexokinase, ADP, EDTA, $MgCl_2$

NH_4Cl (mM)	0	20
ΔO (μmoles)	7.0	5.5
Δest.P (μmoles)	12.2	9.1
P:O	1.74	1.64
Δest.P (corr.*)	9.6	9.1
Decrease in Δest.P(-Δ~)	0.5	
ΔGlu	-4.69	-3.89
ΔAsp	+5.02	7.12
Δ(Glu + Asp)	0.33	3.23
Δ(Glu + Asp):-Δ~	6.5	

* For difference between ΔO with and without NH_4Cl

Fig. 10

Table VIII shows that the value of n for aspartate synthesis can greatly exceed 1, i.e. it appears that ~ acts catalytically rather than stoichiometrically, in this reaction. The fact that some energy is necessary for the synthesis of aspartate is the reason why the value for n with succinate as substrate should be corrected upwards.

TABLE IX

ISOCITRATE AS HYDROGEN DONOR: EFFECT OF INHIBITORS AND P_i ACCEPTOR

Reaction mixture contained α-oxoglutarate, NH_4Cl, arsenite, isocitrate and P_i.

Additions	ΔGlutamate (μmoles)
None	2.2
Oligomycin	2.1
Antimycin	2.0
Amytal	1.9
Dinitrophenol	2.0
Glucose-hexokinase	1.9

Not all nicotinamide nucleotide-linked oxido-reductions require energy. Table IX shows that the reduction of α-ketoglutarate (+ NH_3) by isocitrate does behave according to the text-books, in that it is not inhibited by inhibitors of the oxidation of the reduced nicotinamide nucleotide.

This result and other indications which there is no time to mention brought us back to a proposal suggested some years earlier by Klingenberg (7, 33) that glutamate dehydrogenase is NADP-specific in mitochondria. This suggested that it was the reduction of NADP by NADH formed by the malate dehydrogenase which required the energy. Estabrook and Nissley (34) reported last summer that ATP was required for the reduction of $NADP^+$ by β-hydroxybutyrate, and Danielson and Ernster (35) have recently in elegant experiments provided direct evidence that the reduction of NADP by NADH requires energy.

$$NADH + NADP{\sim}I \rightleftharpoons NADPH + NAD{\sim}I$$

$$NADP^+ + NAD{\sim}I \rightleftharpoons NADP{\sim}I + NAD^+$$

$$\underline{\text{Sum} \quad NADH + NADP^+ \rightleftharpoons NAD^+ + NADPH}$$

Fig. II

TABLE X

INCREASE IN TPNH AND THE DECREASE IN TPN~I ON THE ADDITION
OF DPNH IN THE PRESENCE OF KCN

(EXPERIMENT 161) ADDITION	TPN	TPNH	μmoles/gm protein TOTAL TPN + TPNH	TPN~I	DPN
NONE	0.21	1.70	1.91	2.95	1.27
KCN 0.002 M	0.22	1.70	1.92	2.94	1.13
KCN + DNP 16 μM	0.56	1.60	2.12	2.74	1.13
KCN + DPNH 21.6 mμmoles	0.30	3.58	3.88	0.98	1.90*
DNP 16 μM	4.86	0.00	4.86	--	--

1. TPN + DPNH $\rightleftharpoons$ TPNH + DPN
2. TPN~I + DPNH $\rightleftharpoons$ TPNH + DPN~I

Purvis slide 3/23/59

Whatever the detailed mechanism, surely the energy requirement for these reactions strongly suggests the involvement of some sort of high-energy form of nicotinamide nucleotide. Indeed, all our results and those of Klingenberg, Estabrook and Ernster are consistent with a reaction scheme proposed by Purvis in the spring of 1959 at the American Chemical Society meeting (Fig. 10). Table X is a copy of a slide which he showed.

The addition of DPNH caused more than a doubling of the TPNH concentration, without any significant decline in TPN content (compare lines 2 and 4). The increase in DPN was much less than the increase in TPNH, part of that is accounted for by DPN contaminating the DPNH. The conclusion here is that most of the TPNH was derived from another TPN compound, indicated as TPN~I, and most of the DPN disappears as DPN~I.

Low (25, 26) made the valuable contribution that added NAD^+ could be reduced by succinate in the presence of beef-heart mitochondrial preparations, ATP and an inhibitor of cyrochrome oxidase. Mr. van Dam and Dr. Haas, in unpublished work in our laboratory, have now found that an alkaline extract of such a reaction mixture contains, besides NADH, a species of NAD already described by Hilvers (36) from our group as an intermediate in the reduction of NAD by glyceraldehyde catalysed by glyceraldehydephosphate dehydrogenase - namely an alkaline-stable NAD. Ordinary NAD is completely decomposed in boiling 0.1 M $NaHCO_3$ solution with a half-life of about 14 seconds; the alkaline-stable NAD has a half life of about 100 minutes under the same conditions. At the same time the alkaline extract, after addition of acetaldehyde and alcohol dehydrogenase, showed an absorption maximum at 322 mµ. The relationship between this compound and that described by Griffiths (37, 38) is not yet clear. In fact, these data are very preliminary, and although all the controls we can think of have been done, maybe we shall think up some more in the next few weeks. But we feel hopeful.

The energy-linked reduction of NAD^+ by succinate provides a method of studying the first oxidative phosphorylation step separately from the others - of course in reverse. It is also possible to study the reaction in the forward direction. The oxidation of NADH by fumarate which I showed directly in 1950 (ref. 39) - it was foreshadowed by Green's (40) finding of the oxidation of β-hydroxybutyrate by fumarate - is less interesting than reduction of NAD by succinate, because it runs thermodynamically downhill, although the demonstration that the succinate and DPNH oxidase system could be linked had a certain interest. Dr. Haas has recently shown that the oxidation of NADH by fumarate, catalysed by beef-heart mitochondria, can be coupled to the synthesis of ATP with P:O ratios of 0.7-0.8. This will surprise no-one, but it does not seem to have been done before.

Thus, it has been demonstrated from both sides that the reaction is

Succinate + NAD^+ + 1∼ $\rightleftharpoons$ Fumarate + NADH

REFERENCES

1. Bücher, T., Biochim. Biophys. Acta, $\underline{1}$, 292 (1947).

2. Chance, B., in O.H. Gaebler (Editor) Enzymes: Units of Biological Structure and Function, Academic Press, Inc., N.Y., 1956, p. 447.

3. Chance, B. and Hollunger, G., Fed. Proc., $\underline{16}$, 163 (1957).

4. Chance, B. and Hollunger, G., Nature, $\underline{185}$, 666 (1960).

5. Chance, B. and Hollunger, G., J. Biol. Chem., $\underline{236}$; 1534, 1555, 1562, 1577 (1961).

6. Chance, B., J. Biol. Chem., $\underline{236}$; 1544, 1569 (1961).

7. Klingenberg, M. and Slenczka, W., Biochem. Z., $\underline{331}$, 486 (1959).

8. Klingenberg, M., Slenczka, W. and Ritt, E., Biochem. Z., 332, 47 (1959).
9. Ernster, L., Proc. IUB/IUBS Symp. on Biological Structure and Function, Stockholm, 1960, Vol. 2, Academic Press, N.Y., 1961, p. 139.
10. Ernster, L., Symp. on Intracellular Respiration: Phosphorylating and Non-phosphorylating Reactions, Proc. 5th Intern. Congr. Biochem., Moscow, 1961, Vol. 5, Pergamon Press, London, 1963, p. 115.
11. Snoswell, A.M., Biochim. Biophys. Acta, 52, 216 (1961).
12. Snoswell, A.M., Biochim. Biophys. Acta, 60, 143 (1962).
13. Penefsky, H.S., Biochim. Biophys. Acta, 58, 619 (1962).
14. Slater, E.C., Tager, J.M. and Snoswell, A.M., Biochim. Biophys. Acta, 56, 177 (1962).
15. Tager, J.M., Howland, J.L. and Slater, E.C., Biochim. Biophys. Acta, 58, 616 (1962).
16. Slater, E.C. and Tager, J.M., Fed. Proc. 22, 653 (1963).
17. Packer, L. and Denton, M.D., Fed. Proc., 21, 53 (1962).
18. Tager, J.M., Howland, J.L., Slater, E.C. and Snoswell, A.M., Biochim. Biophys. Acta, in press.
19. Packer, L., Fed. Proc., 20, 1327 (1962).
20. Packer, L., J. Biol. Chem., 237, 1327 (1962).
21. Löw, H. and Vallin, I., Biochem. Biophys. Res. Commun., 9, 307 (1962).
22. Chance, B. and Fugmann, U., Biochem. Biophys. Res. Commun. 4, 317 (1961).
23. Slater, E.C., Nature, 172, 975 (1953).
24. Klingenberg, M. and Schollmeyer, P., Symp. on Intracellular Respiration: Phosphorylating and Non-phosphorylating Reactions, Proc. 5th Intern. Congr. Biochem., Moscow, 1961, Vol. 5, Pergamon Press, London (1963), p. 46.
25. Löw, H., Kreuger, H. and Ziegler, D.M., Biochem. Biophys. Res. Commun., 5, 231 (1961).
26. Löw, H. and Vallin, I., Biochim. Biophys. Acta, 69, 361 (1963).
27. Krebs, H.A., Biochem. J., 80, 225 (1961).
28. Purvis, J.L., Nature, 182, 711 (1958).

29. Purvis, J.L., Biochim. Biophys. Acta, 38, 435 (1960).
30. Slater, E.C., Bailie, M. and Bouman, J., Proc. IUB/IUBS Symp. Biol. Structure and Function, Stockholm, 1960, Vol. 2, Academic Press, N.Y., 1961, p. 207.
31. Hulsmann, W.C., Over het mechanisme van de ademhalingskten-phosphorylering, M.D. Thesis, Klein Offset Drukkerij, Amsterdam, 1958.
32. Tager, J.M., Biochem. J., 84, 648 (1962).
33. Klingenberg, M., Symp. über Redoxfunktionen cytoplasmatischer Strukturen, Gemeinsame Tagung der deutschen Gesellschaft fur physiologische Chemie und der osterreichischen biochemischen Gesellschaft, Wein, 1962, p. 163.
34. Estabrook, R.W. and Nissley, S.P., Symp. über den Mechanismus der Regulation des Zellstoffwechsels, Rottach-Egern, 1962.
35. Danielson, L. and Ernster, L., Biochem Biophys. Res. Commun., 10, 91 (1963).
36. Hilvers, A.G. and Weenen, J.H.M., Biochim. Biophys. Acta, 58, 380 (1962).
37. Griffiths, D.E. and Chaplain, R.A., Biochem. Biophys. Res. Commun., 8, 497 (1962).
38. Griffiths, D.E. and Chaplain, R.A., Biochem. Biophys. Res. Commun., 8, 501 (1962).
39. Slater, E.C., Biochem. J., 46, 484 (1950).
40. Dewan, J.G. and Green, D.E., Biochem. J., 31, 1074 (1937).

DISCUSSION

Sanadi: We find that the ADP to DPNH ratio in the oxidation of DPNH by fumarate is roughly 0.7 to 0.9 in these experiments; the average comes out to about 0.8. This is at saturating levels of added coupling factor; it depends very much on the addition of the coupling factor.

Slater: Dr. Haas did his experiments in beef heart mitochondria, which are sufficiently leaky so that NADH can be oxidized. The P_i:DPNH ratio was 0.7 to 0.8, with susceptibility to inhibitors which you would expect from the reaction. But there is perhaps one significant difference: Antimycin increased the P:O ratio by 0.1.

Löw: May I ask what is the rate of electron transfer between DPNH and fumarate?

Slater: It is a fairly slow reaction. I don't know that I can give you any rates in these mitochondrial preparations. I know what it was in the old Keilin and Hartree preparation, about 1.5-2.0 % of the NADH oxidase activity.

Estabrook: I wonder if you would comment further on the stability of this alkaline-stable DPN. Following Klingenberg alcoholic KOH method, we do not see this. Pressman and Baessler have been doing the acid ferricyanide method and do not see it. Is this compound unstable in more alkaline solution? In other words, your Harkness preparation is pH 10 we are working around pH 13 to 14.

Slater: We do ours at boiling temperature; you do yours at room temperature in absolute alcohol?

Estabrook: Yes.

Slater: Well, that is different from your original preparation, which contained no absolute alcohol, isn't it?

Klingenberg: We changed three years ago to the alcoholic KOH method because we consistently find more DPNH and TPNH with alcoholic extraction (1). We attributed this difference to incomplete extraction by aqueous alkali of the DPNH which may be in the liquid phase.

Slater: We haven't compared the two methods in the particular reaction where we get the alkaline-stable DPN with added DPN.

ENERGY-LINKED FUNCTIONS OF MITOCHONDRIA

I don't think that Hilvers has done the comparison with the phospho-glyceraldehyde dehydrogenase system. He has certainly tested the stability of the alkali-stable DPN that is formed in the glyceraldehyde-phosphate dehydrogenase; it is quite stable at higher pH's.

Estabrook: My understanding of Dr. Griffiths is that his phosphate compound is not alkaline-stable, i.e., it is alkaline-labile, so that there is no relationship between these two derivatives of pyridine nucleotide.

Griffiths: The product of alkaline extraction may not react as NAD or NADH with ADH, i.e., the reaction product may be an NAD or NADH derivative.

Estabrook: Let me get this straight: you are saying that Dr. Slater has your broken-down compound - your compound broken down?

Slater: Was he?

Griffiths: No, I think that the extra DPN described by Purvis (2) could be a mixture of the phosphorylated compound, or its precursor which Dr. Slater says is X $\sim$ I, or a reaction product - none of which react readily in the ADH assay.

Chance: The stoichiometry determinations could be complicated by differences in the nature of the reaction product. One would expect a high efficiency in the diagram which you showed (Fig.12) because the oxidation-reduction potential of the product is that of glutamate and α ketoglutarate, not of DPNH. Also, the reaction is "pulled" by glutamate synthesis; as soon as NADH is formed, it is converted to NAD. But in the absence of a hydrogen acceptor such as α ketoglutarate, the product itself and its thermodynamic potential may differ. Perhaps another molecule of A $\sim$ I is used to react with NADH to convert it into a compound at a lower potential. I certainly agree that there is a difference in our results and, in fact, I think it is very important to establish that there is a reaction in which NADH is formed at the cost of one high-energy phosphate bond. But when we use ATP as the energy source and do not use a hydrogen acceptor, it could "cost" more to get the reaction product. I would suggest that there are other experimental conditions where you can make it a "more expensive" reaction product.

Slater: I completely agree that it is possible that the subsequent reaction of NADH could form this type of compound, but I think our stoichiometry shows that it would not be an intermediate in NADH formation.

Griffiths: What I have to say, I think, bears directly on Professor Slater's work and on the systems studied by Dr. Low, i.e., that energy-dependent reduction of DPN by succinate and I shall confine my remarks to the evidence for a phosphorylated derivative of DPN or DPNH in this reaction. Our interest in this possibility was first aroused by the fact that we saw a compound absorbing at around 320 mμ with the properties of an intermediate in this reaction, and also be some peculiarities in the assay of inorganic phosphate in the presence of DPN.

On incubating mitochondria with DPN and inorganic phosphate at 0° for about three minutes - essentially the conditions used by Purvis in the demonstration of extra DPN - we were able to demonstrate the formation of a phosphorylated derivative of DPN which is a compound containing equimolar amounts of phosphate and DPN. The formation of this compound is inhibited by dinitrophenol and by Amytal. The reaction is also inhibited by Antimycin, which indicates that electron transport is necessary for the accumulation of this compound it is not, however, inhibited by oligomycin. This compound can transphosphorylate to ADP, giving rise to ATP and DPNH in an oligomycin-sensitive reaction. The yield of DPNH is such that two moles of this compound give rise to one mole of DPNH, the other reaction product being a mixture of compounds which we believe are DPNH derivatives, one of which can be identified enzymatically as the compound DPNH-X first described by Edwin Krebs (3), and the other as a compound absorbing at 295 mμ which bears some relation to the acid modification products of DPNH. The following are structures for DPNH-X and the acid modification products of DPNH as written by Burton and Kaplan (4).

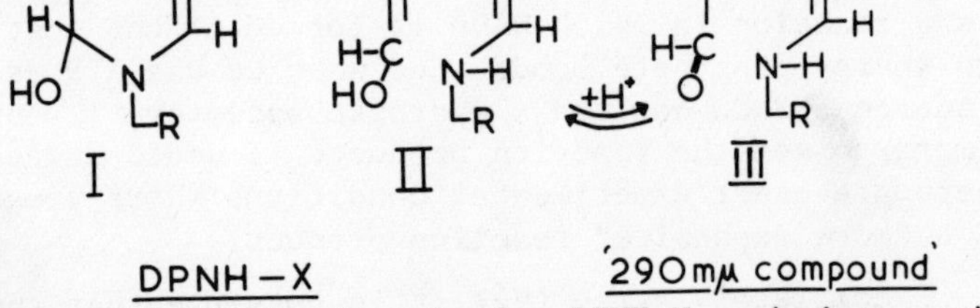

DPNH—X   290 mμ compound

Fig. 1. 1. The Krebs formula for DPNH-X, a hydroxy substituent in the nicotinamide ring of DPN in the 6-position. II. The ring open form as described by Burton and Kaplan. III. The final acid modification product which is the ketone form.

One of the reaction products is the DPNH-X compound, identified by an enzymatic reaction described by Krebs (3) as an ATP-dependent generation of DPNH by an enzyme system from yeast. During the breakdown of one mole of the phosphorylated compound by submitochondrial particles, 0.5 mole of DPNH and 0.1 mole of DPNH-X are formed; the remainder is accounted for by the compound absorbing at 295 mμ. On the basis of these results, we suggest the following tentative structure for this phosphorylated derivative of DPN:

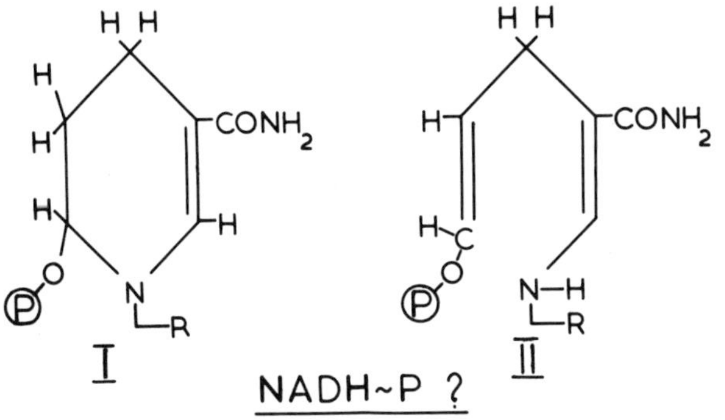

Fig. 2. I. The phosphorylated derivative of DPNH-X. II. The ring-open form.

We believe that this compound satisfies the requirements for an intermediate in the energy-linked reduction of DPN by succinate. First of all, it is formed on incubation with succinate; transphosphorylation to ADP or to water generates DPNH. It could be the direct intermediate, or an intermediate in equilibrium with the non-phosphorylated derivative DPN∿I or DPNH∿I, whatever the structure of that compound may be. We believe that the dinitrophenol sensitivity and the Amytal sensitivity of the formation of this compound, and the oligomycin sensitivity of its transphosphorylation to ADP, characterizes it as an intermediate in this reaction. The formation of DPNH from this compound is not oligomycin-sensitive, which fits in with the known properties of oligomycin.

<u>Slater</u>: This is very interesting indeed, Dr. Griffiths, very exciting work. I would like to ask one question about this DPNH-X which you identify: how do you exclude that you are not getting DPNH-X formed during the working-up of your product? DPNH is very unstable; even at pH 7 you get formation of the acid-modification product from DPNH.

Griffiths: This is the reaction product on incubation of DPN $\sim$P with submitochondrial particles in the presence or absence of ADP. One mole of DPNH is formed from two moles of the phosphorylated compound, and in the reaction mixture, which clarified by centrifugations, we can demonstrate the presence of DPNH-X by the enzyme reaction described by Krebs.

Slater: What is the pH?

Griffiths: Neutral, pH 7.2.

Estabrook: If you get DPNH from your DPNH $\sim$P, might the DPNH-X formation be a subsequent reaction from the DPNH?

Griffiths: I don't think so; consistently we have had this finding of one mole of DPNH per two moles of phosphorylated compound. The amount of DPNH-X is variable, but it is usually 10-20 % of the remaining reaction product. The main reaction product is the compound absorbing at 295 mμ, which could well be an acid degradation product of DPNH.

Mildvan: Of the two possible structures which you propose for this compound, one - the open-ring form - is an enol phosphate and hence a "high-energy" compound. Is the other - the closed-ring form - also a "high-energy" compound, and, if not, might it not be ruled out as a possible intermediate in oxidative phosphorylation?

Griffiths: I really don't know what the possibilities are.

Hess: Is there any evidence for this compound in mitochondria. At low temperatures one should be able to detect it.

Griffiths: The compound is a DPNH derivative, non-fluorescent in the visible, much like DPNH-X of triose phosphate dehydrogenase. It has no 340 mμ absorption, but an increased absorption at 280-290 mμ.

Chance: We have made spectrophotometric measurements at these wave-lengths but do not yet find this compound.

Estabrook: After Dr. Chance and I returned from talking with you last fall, we discussed the extent of absorbancy at 260 mμ. You had indicated that the extinction of your compound is about 8, whereas we know that the adenine ring has an extinction of about 12.

Griffiths: The initial absorption is low, and when the compound decomposes, the extinction at 260-265 mμ is about double. I cannot explain why the initial absorption is low, but a possible explanation is its fluorescence, which would explain the amount of absorption.

ENERGY-LINKED FUNCTIONS OF MITOCHONDRIA

Pressman: What do you mean by low initial absorption?

Griffiths: There is an apparent doubling of the absorption when the compound decomposes, and then you can account for 80 % of the absorption as DPN, by the alcohol dehydrogenase assay*.

Estabrook: Many of us tread a very fine line near secondary reactions, and I noticed that you were quite careful to say that this was the intermediate in the energy-linked reduction of succinate. Are you implying that this may not be, or definitely is not, the intermediate during the normal DPN-linked substrate oxidations?

Griffiths: I don't imply that.

Estabrook: Do you have any evidence to support the case that this is an intermediate in malate oxidation?

Griffiths: In the presence of some DPN-linked substrates (α ketoglutarate and β hydroxybutyrate) I have been able to demonstrate the formation of trace amounts of this compound, but have been unable to demonstrate its formation with pyruvate, malate or glutamate. I don't regard this as evidence against the participation of this compound as an intermediate in the oxidative phosphorylation of DPN-linked substrates, but merely as evidence of the fact that I haven't been able to demonstrate its formation. I believe that in the presence of DPNH, the turnover of this compound is so rapid that the chances of isolating or demonstrating it are extremely small.

Pressman: In the literature there is strong indication that DPNH-X forms primarily under enzymatic conditions (3), but we reported at the Federation Meetings several years ago (6) that we could identify it as the early acid breakdown product in DPNH, and we followed its formation by separating it chromatographically from the compound referred to by Krebs as the "primary acid-breakdown product." This favors the possibility that DPNH-X could actually arise as a chemical degradation product rather than necessarily as the product of an enzymatic reaction.

Griffiths: This is a possibility; my only intention in presenting the evidence for DPNH-X formation is as accessory evidence for the fact that this compound has the properties of DPNH $\sim$ P rather than of DPN $\sim$ P; that this is a compound from which you get DPNH and a DPNH derivative. How the DPNH arises, I don't know; it might well be a non-enzymatic dismutation.

---

* Note added in proof: In a recent paper (5), V.P. Skulachev suggests that the low initial absorption is due to reduction of the adenine ring.

<u>Pressman</u>: When your two moles of DPNH$\sim$P are discharged by the ADP, do you get one mole or two moles of ATP?

<u>Griffiths</u>: I get two moles of ATP.

<u>Green</u>: I was wondering whether we could possibly look at it in another way: there is a high-energy compound of DPNH which does not involve phosphate. There are two ways of phosphorylating this high-energy intermediate, both of which may go under your conditions. If the phosphorylated intermediate is in equilibrium with the original high-energy compound, the question of what is the real intermediate isn't a very meaningful one, because they are all in equilibrium.

If there is a series of equilibria, then the question of what is the true intermediate is an academic one; they are all in equilibrium if there are two ways of phosphorylizing the first product. At least I like to think that.

REFERENCES

1. Klingenberg, M., in Methoden der Enzymatischen Analyse. Verlag Chem., 1962, p. 531.
2. Purvis, J.L., Nature, <u>182</u>, 711 (1958).
3. Rafter, G.W., Chaykin, S. and Krebs, E.G., J. Biol. Chem. <u>208</u>, 799 (1954).
4. Burton, R.M. and Kaplan, N., Arch. Biochem. Biophys., <u>101</u>, 139 (1963).
5. Skulachev, V.P., Nature, <u>198</u>, 444 (1963).
6. Pressman, B.C., Fed. Proc., <u>17</u>, 291 (1958).

MORPHOLOGICAL AND FUNCTIONAL ASPECTS OF PYRIDINE NUCLEOTIDE
REACTIONS IN MITOCHONDRIA
Martin Klingenberg
Physiologisch-Chemisches Institut
der Universitat Marburg, Germany

A theory of the organisation of mitochondria is proposed with specific relation to the cristae of mitochondria in order to comprise in a new way various, apparently heterogeneous data on energy linked functions in mitochondria.

The major structural feature of mitochondria are the internal membranes, the cristae. The structurally bound enzymes are generally assumed to be associated with these membrane structures. This has been placed on a more quantitative basis, by comparing the content of cytochromes and internal membrane of mitochondria in various organs (Vogell, Klingenberg, 1963). In contrast, the soluble components are generally assumed to be located mainly in the matrix space. They are there confined by the outer membrane of the mitochondria from diffusion to the surrounding medium.

The "dynamic organisation" of hydrogen pathways: Outlines of the theory. However, a series of considerations lead us to assume that many of the internal components found in intact isolated mitochondria such as pyridine- and adenine nucleotides as well as some substrates, and metal cations are located in a separate intramitochondrial space. This space may be identified with the "intracristae" space which is the morphologically defined subcompartment in mitochondria.

The following data lead to this assumption: 1. Nucleotides and substrates diffuse freely to a part of the mitochondrial space (Pfaff, 1963). The internally "bound" substances exchange only slowly with the added nucleotides or substrates (Pfaff, 1963). 2. The internally bound pyridine- and adenine nucleotides as well as substrates appear to be largely in an inhibited state, which does not permit these substances to react with the enzymes specific for them in hydrogen or phosphate transfer (Jacobs, Goebell, 1963). 3. The cristae content and DPN content go parallel when comparing, for example, liver and heart mitochondria (Vogell, Klingenberg, 1962). 4. DPN and ubiquinone in mitochondria have a similar redox behaviour in particular with respect to an energy control of their redox state (Szarkowska, Klingenberg, 1963, Chance, 1961). Since it is highly probable that ubiquinone is contained in the lipid phase of the cristae, also the mitochondrial DPN may be in a close spatial relation to the cristae.

It is further proposed that this compartmentation of nucleotides is "dynamic". It is not a fixed compartmentation but changes with the metabolic state of the mitochondria. It appears suggestive that this "dynamic organisation" plays a major role in the phosphorylation reactions.

The dynamic compartmentation is related and controlled by transport phenomena. It has been known for some time that there are transport phenomena across the membrane structures of mitochondria as well as through other membranes.

It is proposed that the energy linked directional reactions which lead to transport phenomena are taking place in the membranes of the cristae. Substances are transported from the matrix to the inner-cristae space. Up to now, transport phenomena in mitochondria have been considered to take place mainly across the outer membrane into the matrix space (cf. Lehninger 1962). It appears to us improbable that only the outer membrane is responsible for mitochondrial transport processes, since the transport reactions can reach a high activity and thus are probably connected to all respiratory chain complexes in the mitochondria. The asymmetry required for these transport processes is given morphologically only across a "unit" membrane of the cristae and not across the whole double membrane.

<u>Examples of energy controlled hydrogen transfer</u>. The validity of the theory may be illustrated best in considering energy linked hydrogen transfer reactions in mitochondria. This has been described for various reactions, which appear to be dependent on energy supply from the phosphorylation system although they do not necessarily require energy. These reactions are designated as "energy controlled" to differentiate them from energy dependent reactions where the energy is required on thermodynamic grounds. Thus, there is the energy controlled transhydrogenase reaction (Klingenberg 1960, Klingenberg et al. 1961, Estabrook et al. 1962, Danielson et al. 1963), the energy controlled hydrogen exchange from malate to ketoglutarate + NH_3 (Klingenberg 1962, Tager 1962), and the energy linked reductions of flavoprotein (Klingenberg et al.

ENERGY-LINKED FUNCTIONS OF MITOCHONDRIA

1961) and ubiquinone. The energy requirement becomes apparent when these reactions take place under anaerobic conditions and there is no energy supply from oxidative phosphorylation.

Hydrogen transfer between the malate and glutamate dehydrogenases. As an example, the ATP controlled hydrogen exchange between malate and ketoglutarate + NH_3, as described here one year ago and also communicated by Tager (1962), might be analyzed for their thermodynamics and kinetics. As illustrated briefly in the scheme of Fig. 5, the addition of malate and ketoglutarate + NH_3 leads to the production of glutamate, which can further produce aspartate by transamination. It was found that glutamate and aspartate are formed to a large extent under anaerobic conditions only when ATP is supplied. The thermodynamic analysis of this reaction showed that the effect of ATP consists in facilitating the equilibration of the system. ATP does not invest energy into increased redox potential differences of the substrates as, for example, in the succinate linked acetoacetate or ketoglutarate + NH_3 reduction.

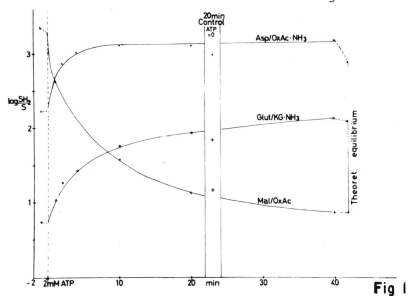

Fig 1

Time dependence of the redox ratios in the malate → ketoglutarate + NH_3 system. Liver mitochondria anaerobically (N_2) incubated with 1.5 mM l-malate, 1.5 mM α-ketoglutarate, 8 mM NH_4Cl, 0.6 mM AsO_3 in 0.25 mM Saccharose- 1 mM EDTA medium at 28° C, pH 7.2. Addition of 2 mM ATP at time zero.

This is demonstrated in Fig. 1, where the time dependence of the redox ratios of the participating substrate coupler are shown. It can be deduced from this presentation that the transamination reaction between glutamate and aspartate is from the beginning near equilibrium whereas the pyridine nucleotide linked redox reactions between the two substrate couples malate/oxalacetate and glutamate/ketoglutarate + NH_3 approach equilibrium only after some time. In the absence of ATP the redox couples do not approach equilibrium as shown in the controls. That ATP acts by reversal of the energy transfer reactions in the oxidative phosphorylation system is suggested by the inhibitory effects of uncouplers (dinitrophenol and oligomycin (Klingenberg 1962).

TABLE I

The Reaction Malate + $NH_3 \rightarrow$ Aspartate in Mitochondria
Under the Influence of
Ketoglutarate (KG) and Glutamate (Glut)

Additions	Aspartate mM
Malate + NH_3	
... + KG	0.13
... + Glut	0.65
... + KG + ATP	0.97
... + Glut + ATP	1.36
... + Glut + DNP	0.50

Anaerobic (N_2) incubation of liver mitochondria, in saccharose - EDTA - medium, pH 7.2, 15 min at $25°C$.

We may turn now to the redox state of the mitochondrial pyridine nucleotides during this hydrogen transfer. It has been shown (Klingenberg and Schollmeyer 1961) that with ketoglutarate + NH_3 in the anaerobic state only a small part of the pyridine nucleotides is oxidized, although a rapid complete oxidation can be expected through the mitochondrial glutamate dehydrogenase. The oxidation is, however, largely increased on addition of malate (cf. Fig. 4, 6). If further ATP is added, the pyridine nucleotides become reduced, in particular, in the presence of hydroxybutyrate. This serves to briefly illustrate that, under the influence of malate and ketoglutarate + NH_3, pyridine nucleotides can become extremely highly oxidized and that ATP can partially prevent this oxidation.

It is suggested that it is the highly oxidized state of the pyridine nucleotides which prevents equilibration between the malate/oxaloacetate and glutamate/ketoglutarate + NH_3 system. Thus the overall reaction, i.e., the formation of aspartate, should proceed without an energy supply when the pyridine nucleotides remain reduced. This can, in fact, be shown by replacing the "oxidizing" substrate, ketoglutarate with a "reducing" substrate, such as glutamate. Under these conditions aspartate is also formed in large amounts in the absence of ATP (Table I). This reaction is further stimulated by ATP, and only partially inhibited by dinitrophenol. Thus it does not have the characteristics of the ketoglutarate system. The ATP independent aspartate formation was further determined as a function of the ratio of glutamate/ketoglutarate (Fig. 2). At ratios of approximately 1 the aspartate formation increased several fold until a maximum was observed at a glutamate/ketoglutarate ratio of approximately 5. At a ratio of infinity, i.e., when no ketoglutarate other than that arising from glutamate is present, aspartate formation is again decreased.

The oxaloacetate level may be regarded as indicating the degree of oxidation of the DPN since it is in equilibrium with the malate dehydrogenase. It decreases inversely to the aspartate level. The inhibition of aspartate formation at a ratio of malate/oxaloacetate of approximately 20 (a condition which strongly oxidizes the DPN system) is overcome when, as reflected in higher malate/oxaloacetate, more reduced pyridine nucleotide is provided from the beginning of the incubation by the addition of glutamate. Thus the reduction of pyridine nucleotides can overcome the ATP requirement for the hydrogen transfer between the malate and glutamate dehydrogenases.

On the other hand, the reaction can be inhibited by a too high reducing power of the substrates such as with glutamate/ketoglutarate + NH_3 equalling infinity. This may also offer an explanation for some effects of malate on the hydrogen transfer from ketoglutarate to ketoglutarate + NH_3, i.e., the ketoglutarate "dismutation" (Klingenberg, Schollmeyer, unpublished). In this instance the liver mitochondria have to be incubated anaerobically without arsenite. In this case 1/2 Mole glutamate is formed for 1 Mole ketoglutarate consumed. Interestingly, (cf. Table 2), glutamate is formed to a larger extent only in the presence of malate. ATP alone does not favour the dismutating hydrogen transfer, whereas with both ATP

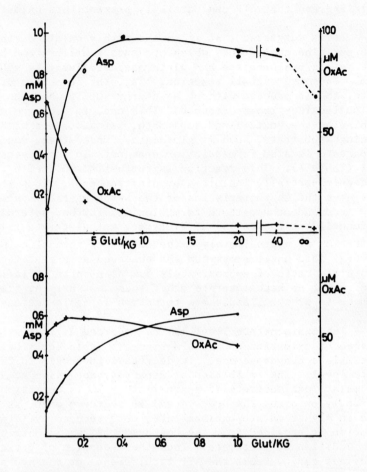

Fig. 2. The reaction malate + $NH_3 \rightarrow$ aspartate in dependence on the ration glutamate/ketoglutarate. Liver mitochondria incubated anaerobically (N_2) with 1.5 mM l-malate, 8 mM NH_4Cl, 0.6 mM AsO_3 in 0.25 mM saccharose - 1 mM EDTA-medium at $28^{o}C$, pH 7.2. Further addition of glutamate + ketoglutarate = 2 mM with varying ratios as indicated.

and malate still more glutamate is formed than with malate alone.

TABLE II

The Reaction Ketoglutarate → Ketoglutarate + NH_3 Under the Influence of Malate and ATP

Addition	Glut	$-\Delta KG$	$\dfrac{-\Delta KG}{Glut}$
	mM		
$KG+NH_3$			
...+Mal	0.35	0.60	1.7
...+ATP	0	0.09	
...+Mal+ATP	0.90	1.12	1.25
$KG+NH_3$			
...+Mal+ATP			
...+3mM Amytal	0.53	0.91	1.7
...+4mM $MgCl_2$	0.60	0.94	1.6
...+6 γ Oligomycin	0.27	0.42	1.6
...+0.1mM DNP	0.42	0.51	1.2

Anaerobic (N_2) incubation of liver mitochondria in saccharose-EDTA-medium, 15 min at $28°$ C.

The ratio Δketoglutarate/glutamate gives the "dismutation coefficient". This permits a differentiation of the dismutation from ketoglutarate reduction by malate or endogenous substrates. Although the effects of ATP are generally inhibited by amytal, magnesium, and oligomycin, a high dismutation ratio is observed which can be attributed to the presence of malate alone. The effect of malate can be related to the oxidation of pyridine nucleotides, which in the presence of ketoglutarate + NH_3 is promoted by malate (cf. below), probably due to the formation of a high oxaloacetate level.

These observations may be summarized as follows: Extensive reduction of pyridine nucleotides, in particular of DPN, may prevent hydrogen transfer. Thus a certain degree of oxidation is required for maximum activity of hydrogen transfer between the dehydrogenases.

<u>The relation of ATP hydrolysis to hydrogen transfer and the redox state of pyridine nucleotide.</u> The role of ATP in these reactions was further investigated by measuring ATP hydrolysis. The analysis showed that ATP is consumed to about 30% excess in comparison to the controls where, in the absence

of malate, the ATP hydrolysis without an extensive hydrogen transfer is measured. The ATP hydrolysis was also followed by simultaneous recordings of pH changes and of pyridine nucleotide absorption as shown in Fig. 3.

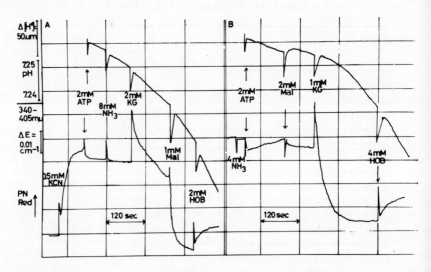

Fig. 3. The rate of ATP hydrolysis in dependence on the redox state of pyridine nucleotides. Simultaneous recording of pH and absorption in a suspension of liver mitochondria. Respiration inhibited with 1 mM KCN.

The rate of ATP hydrolysis increases with the stepwise increased oxidation of pyridine nucleotides by ketoglutarate + NH_3 and then by malate. When ketoglutarate is added last (Fig. 3B), the total oxidation of pyridine nucleotide together with "ATPase" acceleration is seen as one step. Thus we may conclude that the activity of "ATPase" and the degree of oxidation of pyridine nucleotides go parallel, and may be closely related. A similar conclusion was reached by Chance (1962) in studying the ATP hydrolysis with pH recording in mitochondria respiring with succinate.

Interestingly, hydroxybutyrate addition, which causes partial reduction of the pyridine nucleotide, also increases ATPase. In view of our result, that medium reduction of pyridine nucleotide is necessary for maximum hydrogen transfer, one may conclude that ATP serves to maintain reduced pyridine nucleotides and that in this way the malate, ketoglutarate + NH_3 hydrogen transfer utilizes ATP. The reduction of pyridine nucleotide becomes more apparent under the influence of hydroxybutyrate, which amplifies the ATP dependent portion.

ATP controlled DPNH→TPN hydrogen transfer. The ATP induced reduction of pyridine nucleotides in the presence of hydroxybutyrate was shown to reflect mostly a formation of TPNH. This gives direct proof for the previously observed energy control of the redox coupling of the TPN system in mitochondria (Klingenberg and Slenczka 1959). The TPN can be reduced efficiently not only from hydroxybutyrate but also from succinate. The latter, which is a reversal of the flavin-DPN step, has to reduce DPN in an ATP consuming reaction. Both the succinate linked and hydroxybutyrate linked reactions can be differentiated by ATP titration or by the sensitivity to inhibitors and uncouplers (Klingenberg 1963).

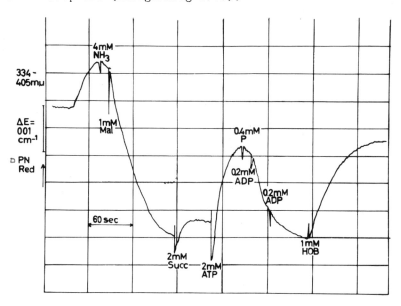

Fig. 4. Differentiation of the phosphorylation potential for the succinate linked and hydroxybutyrate linked TPN reduction. Recording of a suspension of liver mitochondria incubated in saccharose-EDTA-medium at 20° C.

This is further seen in Fig. 4 where in the presence of succinate the ATP effect is first abolished by low phosphate and ADP concentration, and then a reduction with hydroxybutyrate can still be obtained. Thus the transhydrogenation alone (with hydroxybutyrate) is much less sensitive to the ATP potential than reversed hydrogen transfer from succinate.

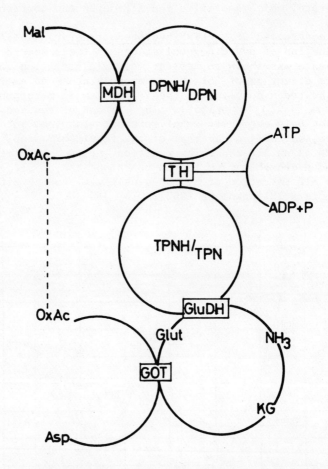

Fig. 5. The possible role of the energy controlled transhydrogenation in the malate → ketoglutarate + NH_3 hydrogen transfer and aspartate formation. TH = transhydrogenase; GluDH = glutamate dehydrogenase; MDH = malate dehydrogenase.

It is feasible that the energy controlled transhydrogenation is the cause for the energy control of hydrogen from malate to ketoglutarate + NH_3. Here a transhydrogenation step is involved if the glutamate dehydrogenase in mitochondria reacts preferentially with the TPNH instead of with the DPNH (Fig. 5, cf. Klingenberg, Pette 1962). However, in this case the velocity of the overall reaction may not be dependent on the redox and binding state of a large part of the TPN system, since the further addition of hydroxybutyrate increases considerably the reduced bound portion of TPNH without influencing the malate ketoglutarate + NH_3 hydrogen transfer (Klingenberg 1962). The hydrogen transfer reaction must therefore be able to work fully with a smaller portion of the TPN system.

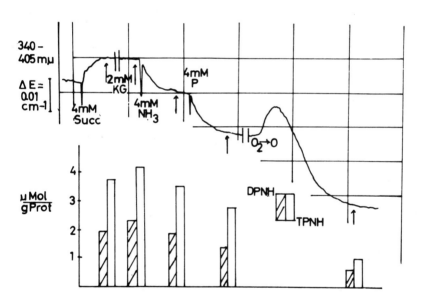

Fig. 6. The effect of phosphate and malate on releasing reduced pyridine nucleotides to oxidation by ketoglutarate + NH_3. Recording in a mitochondrial suspension with simultaneous automatic sampling for enzymatic analysis on DPNH and TPNH. (Klingenberg 1963). Liver mitochondria incubated under N_2 in saccharose-EDTA-medium at 20° C.

Mitochondrial pyridine nucleotides in states of inhibition and release to dehydrogenases. As shown in Fig. 6, part of the DPNH and most of the TPNH resists oxidation by ketoglutarate + NH_3 despite the presence of the DPN and TPN specific glutamate dehydrogenase in mitochondria. Only after further addition of phosphate (which affects more the DPNH) and malate (which affects the TPNH) the pyridine nucleotides are "released" permitting their oxidation via the glutamate dehydrogenase. The subsequent reduction of TPN after ATP addition (here in the presence of hydroxybutyrate) corresponds to a "rebinding" of the TPNH which is then unavailable to oxidation. Thus the role of ATP can be seen in a "binding"-function, which in the presence of hydrogen donors is exerted in particular on the reduced forms of the pyridine nucleotides. The binding or compartmentation of the pyridine nucleotides may be called "energy controlled," since it occurs whenever energy is available whether from ATP under anaerobiosis or directly from the respiratory chain during respiration.

The effect of phosphate may be explained as antagonizing the binding directly by anion exchange, as discussed below, and indirectly by lowering the energy potential. Malate may release the binding indirectly by facilitating oxidation of the reduced pyridine nucleotide, since it provides a high level of oxaloacetate which apparently has a special ability to reach inhibited reduced pyridine nucleotide.

The effects of phosphate on energy controlled substrate-substrate hydrogen transfer are summarized in Table 3. In the malate $\rightarrow$ ketoglutarate + NH_3 system the presence of phosphate increases considerably the aspartate formation in the presence of oxygen when the reaction is controlled by energy supply from respiration. No effect is seen under anaerobic conditions.

In the respiration dependent hydrogen transfer from succinate to DPNH-oxidizing substrates, phosphate also stimulates the reaction two-fold when ketoglutarate + NH_3 serves as acceptor (cf. also Slater et al. 1962). In contrast, phosphate has only a minor influence on the succinate linked acetoacetate reduction (Ernster 1960, Klingenberg and v. Haefen 1963). This shows that the "releasing" action of phosphate is important only for the ketoglutarate + NH_3 reduction and not for acetoacetate reduction. The latter can react also with the "bound" reduced pyridine nucleotides. As illustrated in a scheme below (Fig. 7), this can be interpreted as a different localisation of the dehydrogenases: glutamate dehydrogenase being localised near the surface of the cristae and hydroxybutyrate dehydrogenase in the unit membrane of the cristae.

TABLE III
Influence of Phosphate on Intramitochondrial Hydrogen Transfer

Additions	Aspartate mM	Additions	
1. Malate + NH$_3$ → Aspartate		2. Succinate → Acceptor	
O_2		O_2	Glutamate mM
Malate + NH$_3$ Ketoglutarate	0.175	Succinate+NH$_3$+ Ketoglutarate	0.55
....+4 mM P	0.430	....+4 mM P	0.99
N_2	0.095		Hydroxybutyrate mM
....+4 mM P	0.148	Succinate+ Acetoacetate	1.48
....+2 mM ATP	0.602	....+4 mM P	1.69

Liver mitochondria incubated in saccharose-EDTA-medium
Different preparations for each experimental pair.

The theory of dynamic compartmentation applied to pyridine nucleotide function: The relation between function and dynamic compartmentation of pyridine nucleotides. The general conclusions of this brief experimental survey may be summarized as follows: The pyridine nucleotide function in hydrogen transfer is dependent on a certain medium degree of reduction in the mitochondria. In a high degree of oxidation the pyridine nucleotides are unavailable to dehydrogenases. This state is paralleled by a low energy potential and can result in an irreversible loss of the pyridine nucleotides from the dehydrogenases and from the mitochondria. In a high degree of reduction the pyridine nucleotides also are unavailable to some pyridine nucleotide-linked dehydrogenases due to a binding of the reduced pyridine nucleotide inside the mitochondria in a particular compartment. This binding is supported by a high energy potential.

SCHEME I
Compartments of Pyridine Nucleotides

Location	Intra-Cristae	Cristae Surface	Matrix Extramitoch.
State	inactive with dehydrogenases	active with dehydrogenases	dislocated from dehydrogenases
	DPNH $\underset{PO_4}{\overset{ATP}{\rightleftarrows}}$ DPNH $\updownarrow$ DPNH		$\xrightarrow{PO_4}$ DPN
Enzyme-location	RCDH TH	HOBD MDH	GluDH

RCDH = respiratory chain dehydrogenase
TH = transhydrogenase
HOBD = hydroxybutyrate dehydrogenase
MDH = malate dehydrogenase
GluDH = glutamate dehydrogenase

Thus the pyridine nucleotides are distributed over 3 compartments as described in Scheme I. These functional compartments may find their morphological interpretation according to the lines discussed in the introduction: (1) The matrix space preferred by oxidized pyridine nucleotides, (2) the cristae surface-space with the pyridine nucleotides actively interacting with the dehydrogenases, and (3) the intracristae space preferred by the reduced and bound pyridine nucleotides.* In addition there is assumed a distribution of the dehydrogenases over the 3 compartments, which is tentatively included in Scheme I. In particular this distribution is expected to be ramified on the basis of future results.

<u>Dynamic compartmentation as an active transport dependent process</u>. Active transport processes are regarded to be the basis for the energy controlled compartmentation, as discussed above. In this way the compartmentation can be explained as a dynamic process, which can repeatedly be renewed and abolished. It permits the exchange of nucleotides between the compartments in the metabolic turnover. This is explained in Table IV.

*In discussions Estabrook (1962) also had expressed the view that the DPN reducible by succinate may be localized in the cristae.

TABLE IV

Proposed Cation - Anion Transport Across the Cristae Membrane
Involved in Compartmentation of Pyridine Nucleotides

Metabolic control	Transport processes outer ↔ inner cristae space
DPN reduction ATP hydrolysis	$H^+ \longrightarrow$ $DPNH^{2-} \quad\quad TPNH^{1-} \longrightarrow$ $\longleftarrow$
DPN oxidation ATP formation	$\longleftarrow$ $DPN^- \quad\quad TPN^{3-} \quad H^+$ $\longleftarrow$
	$K^+ \longrightarrow$ $\longleftarrow \quad H^+$ Mg^{++}, Ca^{++} $\longrightarrow$
Exchange Reactions :	$HPO_4^{2-} \longrightarrow$ $DPNH^{2-}, TPNH^{4-}; DPN^-, TPN^{3-}$ $\longleftarrow$

The major driving force may be active H^+-transport, which has been often associated with redox reactions (Davies and Ogston 1950). In the arrangement proposed here H^+ are injected into the inner cristae space. H^+ may be exchanged with K^+ and Mg^{++} or with Mn^{++}, Ca^{++} under unphysiological conditions (Vasington et al. 1962, Brierley et al. 1962, Chappell et al. 1962). The pyridine nucleotides are here regarded as anions $DPNH^{2-}$, etc.), which accompany the active cation transport and are in this way brought into the cristae. This is favored by reduction of the pyridine nucleotides, since the reduced form has 1 more negative charge than the oxidized form. Thus the compartmentation of the pyridine nucleotides is explained to be energy (ATP) dependent.

The energy controlled (dynamic) organisation of hydrogen pathways. The organisation of mitochondrial hydrogen pathways

Membrane transport controlled pathways of pyridine nucleotides in mitochondrial de- and transhydrogenations

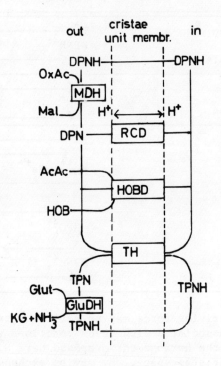

Fig. 7. Proposed organisation of pyridine nucleotide linked pathways across the cristae unit membrane. Pyridine nucleotide compartmentation to the inner space is energy controlled by coupling to active H^+ transport as described in Table 3. The bound DPNH and TPNH is only available to the structurally bound dehydrogenase such as the respiratory chain dehydrogenase (RCDH), the hydroxybutyrate dehydrogenase (HOBDH) and the transhydrogenase (TH). Due to their arrangement in the membrane structure these enzymes may be "directional," i.e., they catalyze preferentially one direction of the reaction, for example the transhydrogenation:

$$DPNH\text{-}in + TPN\text{-}out \longrightarrow DPN\text{-}out + TPNH\text{-}in$$

is regarded as a function of the dynamic compartmentation of pyridine nucleotide superimposed on the fixed compartmentation of the enzymes. The morphological element for this organisation is given by the cristae, which provide the loci for both types of compartmentation.

This is tentatively represented schematically in Fig. 7 for the pyridine nucleotides linked hydrogen pathways: Near the outer surface of the cristae (in the matrix) are localised loosely bound dehydrogenases, such as malate and glutamate dehydrogenases. In the unit membrane structurally bound enzymes such as the respiratory chain and hydroxybutyrate dehydrogenases and the transhydrogenase are located. There these dehydrogenases form "directional enzymes" (Klingenberg 1963, in prep.) where an asymmetric localisation can lead to a preferred direction of the reactions. Thus the respiratory chain dehydrogenases react preferentially with bound DPNH and form directly "bound" DPNH, e.g. from succinate in the energy dependent hydrogen transfer.

This principle may be applied in particular to explain the energy controlled transhydrogenation: Only "bound" DPNH can react with the transhydrogenase to form "bound" TPNH and release DPN. The TPNH formed by transhydrogenation is available to glutamate dehydrogenase only when it is released to the outside by anion exchange etc., as described in Table IV.

SUMMARY

An explanation has been proposed for the energy control of hydrogen transfer of pyridine nucleotides. This involves a newly developed theory of "dynamic organisation" of metabolic pathways in mitochondria. This theory is of a general nature and should comprise also the reactions of other metabolites and nucleotides. The pyridine nucleotides afford one way of introduction to this field.

Thus in conjunction with the concept of "directional enzymes" an explanation of the energy controlled transhydrogenation is given. This may explain as well the energy control of hydrogen transfer between pyridine nucleotide linked substrates.

The cooperation of the energy transfer system of oxidative phosphorylation of active transport processes and of the morphological organisation of mitochondria, as contained in the described theory of dynamic compartmentation, satisfies the postulate (Mitchell 1961, Lehninger 1962) to evaluate these aspects in interpretations of mitochondrial metabolism.

REFERENCES

1. Brierley, G. P., Bachmann, E., and Green, D. E., Proc. Natl. Acad. Sci. U. S., $\underline{48}$, 1928 (1962).

2. Chance, B., Nature, $\underline{195}$, 150 (1962).

3. Chance, B., in A Ciba Foundation Symposium on Quinones in Electron Transport, 1960, Churchill, Ltd., London, 1961, p. 327.

4. Chappell, B., Greville, G. D., and Bicknell, K. E., Biochem. J., $\underline{84}$, 61 (1962).

5. Danielson, L., and Ernster, L., Biochem. Biophys. Res. Comm., $\underline{10}$, 91 (1963).

6. Davies, R. E., and Ogston, A. G., Biochem. J., $\underline{46}$, 324 (1950).

7. Ernster, L., IUB/IUBS Symposium on Biological Structure and Function, Stockholm, 1960, Academic Press, 1961, Vol. II, p. 139.

8. Estabrook, R. W., and Nicholls, P., Wiss. Konf. Rottach-Egern "Funktionelle und Morphologische Organisation der Zelle," 1962, Springer Verlag, 1963, p. 118.

9. Jacobs, H., and Goebell, H., unpublished.

10. Klingenberg, M., and Slenczka, W., Biochem. Z., $\underline{331}$, 334 (1959).

11. Klingenberg, M., II. Mosbacher Kolloquium "Zur Bedeutung der Freien Nukleotide," 1960, Springer Verlag, 1961, p.82.

12. Klingenberg, M., and Bücher, Th., Biochem. Z., $\underline{334}$, 1 (1961).

13. Klingenberg, M., and Schollmeyer, P., Proc. Vth Intern. Congr. Biochem., Moscow, 1961. Symp. on Intracell. Resp., in press.

14. Klingenberg, M., and Pette, D., Biochem. Biophys. Res. Comm., $\underline{7}$, 430 (1962).

15. Klingenberg, M., Symp. Wien 1962, "Redoxfunktionen Cytoplasmatischer Strukturen," p. 163.

16. Klingenberg, M., and v. Haefen, H., Biochem. Z., $\underline{337}$, 120 (1963).

17. Klingenberg, M., Biochem. Z., in preparation.

18. Lehninger, A. L., Physiol. Rev., $\underline{42}$, 467 (1962).

19. Mitchell, P., Nature, 191, 144 (1961).
20. Pfaff, E., unpublished.
21. Slater, E. C., Tager, J. M., and Snoswell, A. M., Biochem. Biophys. Acta, 56, 177 (1962).
22. Szarkowska, L., and Klingenberg, M., Biochem. Z., in press.
23. Tager, J. M., Biochem. J., 84, 64 P (1962).
24. Vasington, F. D., and Murphey, J. V., J. Biol. Chem., 237, 2670 (1962).
25. Vogell, W., and Klingenberg, M., Wiss. Konf. Rottach-Egern "Funktionelle und Morphologische Organisation der Zelle," 1962, Springer Verlag, 1963, p. 59, 64.

DISCUSSION

<u>Slater</u>: I didn't quite follow your mechanism for the effect of phosphate. This requirement for phosphate for all our glutamate dehydrogenase-requiring reactions has been for us a great complication. We found that phosphate was necessary for maximum glutamate synthesis, no matter what the donor - even when the isocitrate was the donor in a reaction which is not energy linked. Since phosphate is not necessary for acetoacetate formation, we just conclude that phosphate has something to do with glutamate dehydrogenase and nothing to do with the reaction we are really studying.

<u>Klingenberg</u>: Our experimental data show that phosphate increases the level of oxidized pyridine nucleotide. We explain this by saying that phosphate is an anion which exchanges with the bound form of pyridine nucleotide and increases the pool of active pyridine nucleotide outside which can exchange with the dehydrogenases.

<u>Slater</u>: Is this specific for glutamate dehydrogenase?

<u>Klingenberg</u>: Yes, the arrangement (cf. Fig. 7 of my paper) was that glutamate dehydrogenase is outside and that the β-hydroxybutyrate dehydrogenase is in the membrane. It does not need phosphate, since β-hydroxybutyrate dehydrogenase can react with the internal DPNH. I think this also makes sense for the proposed role that β-hydroxybutyrate dehydrogenase plays in living cells (1).

<u>Pressman</u>: My question is based on our observation that extramitochondrial adenosine nucleotides permeate mitochondria very freely. By the use of differential labelling techniques, ATP enters the mitochondrially bound pool rapidly and without prior breakdown, whereas we find radioactive DPN does not enter the bound phase except in very small quantities (2). The scheme which you propose implies a common intramitochondrial phase for the free adenosine nucleotides and the free pyridine nucleotides. I wonder if you regard our evidence as consistent with your scheme.

<u>Klingenberg</u>: Yes, there should be a highly oxidized state of the pyridine nucleotides to have communication with the added pyridine nucleotides. Added adenine nucleotides cannot communicate freely with the bound adenine nucleotides, ATP or ADP;

they penetrate into the mitochondria but they equilibrate rather slowly with the endogenous adenine nucleotides.

Green: I would like to know how they permeate. How do you know they get in?

Klingenberg: We make experiments similar to Bartley's with some modifications, and then assay for the ATP that gets in.

Pressman: Then your observations are not consistent with ours. I am surprised that using his technique, Bartley has reported that extra- and intra-mitochondrial pyridine nucleotides equilibrate extensively (3). Our separation technique (4) shows this not to occur at all. So here Bartley's technique seems to show the reverse of what we find, i.e., that the pyridine nucleotides enter the mitochondrially bound phase while we find that they do not, and that adenosine nucleotides are not entering where we find that they do. There must be a serious discrepancy somewhere.

Klingenberg: May I show one experiment which is related to this point?

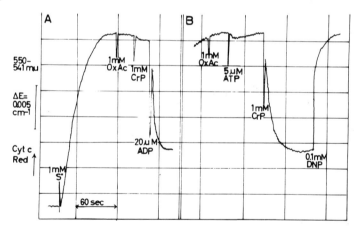

Fig. 1. Creatine phosphate as an energy donor in reversed electron transfer. Heart muscle mitochondria inhibited with sulfide. Oxidation of cytochrome c in energy dependent reversed electron transfer only by the combined addition of Cr-P and ADP (or ATP). Saccharose-EDTA-medium + 4 mM Mg^{++}.

Using mitochondria which contain as internal enzymes creatine kinase, and measuring the energy-dependent oxidation of the cytochromes, we observed that added creatine phosphate alone does not oxidize cytochrome but that after further addition here of 20 μM ADP a large oxidation occurs.

Chance: You do know that the creatine kinase is there?

Klingenberg: Yes. The oxidation is also obtained even with addition of 5 μM ATP when the concentration of endogenous ATP, calculated for the suspension, would give for example 10 μM (we can take ADP or ATP, it doesn't matter). Thus the internal bound adenine nucleotides do not communicate at all with the creatine kinase in the mitochondria. Only external adenine nucleotides even in very small amounts can react in a reversal of oxidative phosphorylation.

REFERENCES

1. Klingenberg, M. and Haefen, H.v., Biochem. Z., 337, 120 (1963).
2. Pressman, B.C., Fed. Proc. 17, 291 (1958).
3. Birt, L.M. and Bartley, W., Biochem. J., 75, 303 (1960).
4. Pressman, B.C., J. Biol. Chem., 232, 967 (1958).

THE INTERACTION OF MITOCHONDRIAL PYRIDINE NUCLEOTIDES

R.W. Estabrook*, F. Hommes **, and J. Gonze

Department of Biophysics and Physical Biochemistry
Johnson Foundation for Medical Physics
University of Pennsylvania, Philadelphia 4, Pennsylvania

The topic of discussion for the first part of this symposium concerns the interplay of the enzymes of oxidative phosphorylation with the reduction of pyridine nucleotides. These are reactions of the type first demonstrated by Chance and Hollunger (1) with their studies of the succinate linked reduction of DPN. More recently it has become apparent that a comparable reaction occurs with the transfer of reducing equivalents from DPNH to TPN during the asymmetric transhydrogenase reaction. Since this presentation bridges the gap between Dr. Klingenberg and Dr. Danielson, it is appropriate to discuss and summarize our own studies related to the latter reaction, i.e., the DPNH reduction of TPN. During these studies the interactions of the endogenous pyridine nucleotides of intact liver mitochondria were investigated. The original program of study was then extended to include measurements of the reaction of pyridine nucleotides added to sonic particles derived from heart muscle sarcosomes.

The realization of a possible involvement of ATP, or the energy derived from ATP, in the operation of the transhydrogenase reaction was revealed in 1961. This was proposed by Klingenberg (2), who had investigated the reduction of the pyridine nucleotides of intact mitochondria, as well as by Estabrook, Fugmann, and Chance (3), who had carried out similar studies with digitonin particles prepared from liver

---

This study was supported in part by a U.S. Public Health Service Grant (RG 9956).
* This work was carried out during the tenure of a U.S. Public Health Service Research Career Development Award (GM-K3-4111).
** Fulbright Scholar 1961-1963. Permanent address: Department of Biochemistry, School of Medicine, University of Nijmegen, The Netherlands.

mitochondria. The need for this hypothesis, at least in our own mind, was to explain the observation of the preponderance of TPN reduced in mitochondria or submitochondrial digitonin particles during β hydroxybutyrate oxidation. These studies were extended by Klingenberg (4), who summarized the in-balance of the DPN-DPNH couple and the TPN-TPNH couple of mitochondria pyridine nucleotides. Figure 1 presents the data of Klingenberg (4), where he employed a number of substrates under various conditions of oxidative phosphorylation. The present studies described here have confirmed many of these observations of Klingenberg, using rather different techniques of analyses as well as different means of monitoring pyridine nucleotide reduction. The only major discrepancy between the two studies is our observation of a more pronounced oxidation of TPNH in State 2B. Attention should be focused on the observation that about 80 per cent of the TPN of mitochondria is reduced under conditions where only 10 per cent of the DPN is reduced. The question arises as to the explanation for this difference in extent of pyridine nucleotide reduction. As described below, this difference may be related to the operation of an energy linked transhydrogenase reaction.

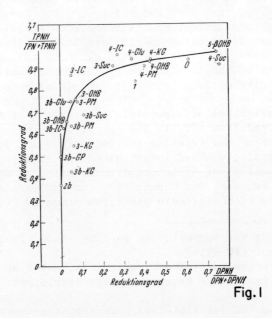

Fig. 1

The percentage reduction of mitochondrial DPN and TPN with various substrates. Data reproduced from Klingenberg (4).

ENERGY-LINKED FUNCTIONS OF MITOCHONDRIA

Effect of Uncouplers

One approach to this problem is illustrated by a series of experiments carried out with intact liver mitochondria employing β hydroxybutyrate as substrate in the presence of an uncoupler of oxidative phosphorylation. Under these conditions the in-balance of pyridine nucleotide reduction was determined in the presence of varying concentrations of octyl dinitrophenol. One such determination is shown in Figure 2.

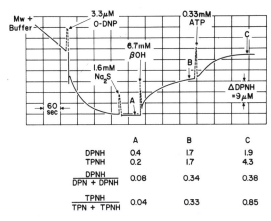

	A	B	C
DPNH	0.4	1.7	1.9
TPNH	0.2	1.7	4.3
$\frac{DPNH}{DPN + DPNH}$	0.08	0.34	0.38
$\frac{TPNH}{TPN + TPNH}$	0.04	0.33	0.85

DPN + DPNH = 5.0 mμMole/ml
TPN + TPNH = 5.1 mμMole/ml

Fig. 2

The effect of octyl dinitrophenol and ATP on the reduction of mitochondrial pyridine nucleotides.

This figure presents a fluorometric recording of pyridine nucleotide reduction, where an increase in fluorescence is indicated by an upward deflection of the tracing, i.e., an increase in fluorescence is synonymous with pyridine nucleotide reduction (cf. Estabrook) (5). When mitochondria are diluted in an isotonic buffer, the pyridine nucleotides are largely reduced due to endogenous substrates. The addition of low levels of octyl dinitrophenol causes a rapid oxidation of the reduced pyridine nucleotides. Sodium sulfide is then added as a terminal inhibitor of the respiratory chain, causing little or no change in the extent of pyridine nucleotides reduced. As indicated in Figure 2, the addition of β hydroxybutyrate as substrate causes a partial reduction of the pyridine nucleotides. The subsequent addition of ATP, however, causes a further increase in fluorescence associated with the reduction of pyridine nucleotide. Of interest is the question which of the pyridine nucleotides are reduced under the various conditions described? To answer this question, samples were removed at the points indicated (A, B and C) and the reaction terminated by addition of the samples to perchloric acid or alcoholic potassium hydroxide. After suitable neutralization, the concentrations of DPN, TPN, DPNH, and TPNH in the samples were determined by the microanalytical fluorometric technique described previously (6). The results of these analyses are summarized in the lower portion of the figure. The analytical results show that mitochondria contain principally DPN and TPN with little DPNH and TPNH when sulfide is added after uncoupler and prior to substrate. The subsequent addition of ATP causes only a small increase in the extent of DPN reduction, and a much larger reduction of the TPN.

The results presented in Figure 2 illustrate three points
1. The concentration of endogenous substrate is decreased to a minimal value upon addition of an uncoupler to mitochondria i.e., no reduction of pyridine nucleotide is observed upon addition of a terminal inhibitor of the respiratory chain.
2. Only a partial reduction of the endogenous pyridine nucleotides of mitochondria is obtained when β hydroxybutyrate is added as a substrate in the presence of an uncoupler of oxidative phosphorylation.
3. The addition of ATP after β hydroxybutyrate establishes a more pronounced reduction of TPN relative to DPN.

The effect of varying concentrations of uncoupler on the extent of pyridine nucleotide reduction was determined in a series of experiments of the type represented in Figure 2. The results are summarized in Figure 3. The dotted line represents the balance of pyridine nucleotide reduction obtained in experiments comparable to that described above for Figure 1. The results of pyridine nucleotide analyses in the presence of various concentrations of uncoupler are represented in Figure 3 by the X points. The concentrations of octyl dinitrophenol present in each experiment are indicated in the associated brackets. These points represent the extent of pyridine nucleotide reduction observed after treatment of mitochondria with octyl dinitrophenol, sulfide, and β hydroxy butyrate, but not ATP (cf. Sample B of Figure 2). If the mitochondrial pyridine nucleotides are initially oxidized by pretreatment with ADP prior to addition of sulfide, the addition of β hydroxybutyrate causes about 55 % reduction of the DPN and 100 % reduction of the TPN. Figure 3 shows that the addition of increasing concentrations of octyl dinitrophenol to the reaction system changes the balance of DPN and TPN reduction. These changes are due mainly to decreases in the extent of TPN reduction with increasing concentration of uncoupler. In each instance the subsequent addition of ATP to the reaction mixture increases the extent of TPN reduction so that nearly 100 % of the TPN is observed.

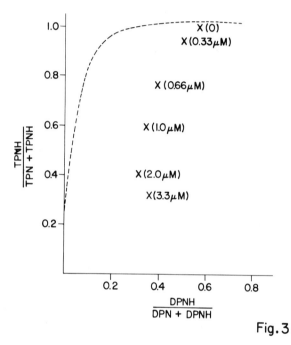

Fig. 3

The extent of pyridine nucleotides reduced by β hydroxybutyrate in the presence of various concentrations of octyl dinitrophenol. The X points represent the results of analyses obtained under conditions similar to Sample B in Figure 2. The concentrations of octyl dinitrophenol added are included in the brackets. The dashed curve represents the balance of pyridine nucleotides reduced under various metabolic conditions (cf. Figure 1).

Effect of Phosphate on the Oxidation of Reduced Pyridine Nucleotides:

A second facet of this study is concerned with the factors influencing DPNH and TPNH oxidation in mitochondria. An

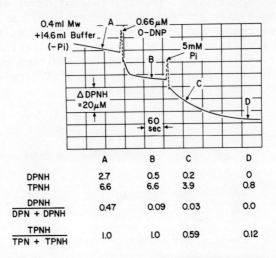

Fig. 4

The requirement for inorganic phosphate in the oxidation of mitochondrial TPNH. Rat liver mitochondria were diluted in a phosphate free buffer and octyl dinitrophenol and inorganic phosphate added as indicated. Samples for analysis were withdrawn at the points A, B, C and D, and the results are summarized in the lower portion of the figure.

example of this type of experiment is presented in Figure 4. It should be noted that the studies described here require rather special conditions of uncoupler concentration, i.e., maximal effects are obtained when suboptimal concentrations of uncoupler are employed. In the experiment illustrated in Figure 4, rat liver mitochondria are diluted in a reaction medium devoid of added inorganic phosphate and the changes in pyridine nucleotide reduction measured fluorometrically. Initially it is observed that the pyridine nucleotides are largely reduced and they are only slowly oxidized. This is shown by the relative constancy of the fluoroescence tracing. The addition of octyl dinitrophenol to such a reaction system causes the oxidation of DPNH but not the oxidation of TPNH. This is indicated by the results of associated analytical measurements summarized in the lower portion of the figure. The subsequent addition of phosphate to the reaction system does cause a decrease in fluorescence associated with the oxidation of the endogenous TPNH. When higher concentrations of uncoupler are added, both DPNH and TPNH are oxidized, presumably due to the activation of a mitochondrial ATPase which liberates sufficient endogenous phosphate to permit the reaction to proceed.

Reaction of Exogenous Pyridine Nucleotides:

The counterpart to studies on the reactions of the endotenous pyridine nucleotides are studies on the interaction of pyridine nucleotides added to submitochondrial particles employing a system comparable to that described by Danielson and Ernster (7). These studies avoid the complex reactions occurring in intact mitochondria by eliminating the question of the compartmentation of mitochondrial pyridine nucleotides (cf. Klingenberg, this symposium).

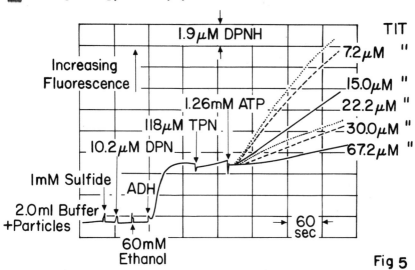

Fig 5

The kinetics of the ATP dependent reduction of TPN with submitochondrial particles. The reaction was measured after addition of varying concentrations of triiodothyronine. Experimental conditions are described in reference (12).

The ATP dependent reduction of TPN by DPNH is illustrated in Figure 5. Particles prepared by sonic treatment of heart muscle mitochondria are suspended in tris buffer containing magnesium chloride. Sulfide, DPN, alcohol, alcohol dehydrogenase, and TPN are added as indicated. As described by Danielson and Ernster (7), no reduction of TPN occurs until ATP is added to the reaction system. Since Danielson and Ernster (7) had questioned the participation of Kaplan's transhydrogenase (8) in this reaction, experiments were desig ed to see if evidence could be obtained for or against the participation of the "classical transhydrogenase" in the ATP dependent reduction of TPN. Ball and Cooper (9) and Stein <u>et al</u> (10) established the inhibition of the classic transhydrogenase by thyroxine and triiodothyronine (TIT). The experiments shown in Figure 5 demonstrate the inhibition of the ATP dependent reduction of TPN by triiodothyronine. Titrations o triiodothyronine concentration versus percentage inhibition are summarized in Figure 6. Three reactions have been measur

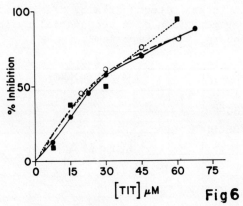

Fig 6

Inhibition of the ATP dependent reduction of pyridine nucleotides by triiodothyronine. Solid circles represent the ATP dependent reduction of TPN by DPNH; solid squares represent the succinate reduction of DPN; and the open circles represen the oxidation of TPNH by DPN. Experimental conditions are de cribed in reference (12).

in the presence of varying concentrations of triiodothyronine These are as follows:

(1) TPN + DPNH $\xrightarrow{\text{ATP}}$ TPNH + DPN

(2) TPNH + DPN $\longrightarrow$ DPNH + TPN

(3) Succinate + DPN $\xrightarrow{\text{ATP}}$ Fumarate + DPNH

As shown in Figure 6 all three reactions have the same sensitivity to triiodothyronine.

A second criterion has been applied to test the participation of the classic transhydrogenase in the ATP dependent reduction of TPN. Kaplan (11) has observed a high (> 20 Kcal) activation energy for the transhydrogenase reaction. Measurements on the influence of temperature on the rate of the three reactions described above, all showed similar high activation energies. On the basis of the similarity of triiodothyronine inhibition as well as the high activation energy of the reaction, it has been concluded (12) that transhydrogenase participates in the ATP dependent reduction of pyridine nucleotides.

Summary:

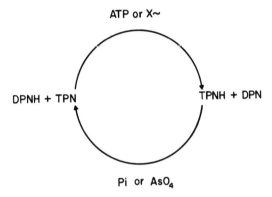

Fig. 7

Schematic representation of the mitochondrial transhydrogenase reaction.

The studies described on the mitochondrial transhydrogenase reaction are summarized schematically in Figure 7. It is apparent that the reaction of DPNH with TPN is facilitated by the addition of ATP. This has been demonstrated directly with the endogenous pyridine nucleotides of intact mitochondria as well as with exogenous pyridine nucleotides added to submitochondrial particles. The exact manner by which ATP influences this reaction is unknown, although Danielson and Ernster (7) have shown that ATP can be replaced by X$\sim$ generated during substrate oxidation. The reversal of this reaction, i.e., the oxidation of TPNH by DPN, is affected by inorganic phosphate. This has been demonstrated by direct measurements of the endogenous pyridine nucleotides of rat liver mitochondria. The reaction scheme of Figure 7 is represented in a cyclic form to exclude the conclusion that ATP is formed during the oxidation of TPNH by DPN. Presently no evidence exists to conclude that ATP synthesis can occur during the transhydrogenase reaction.

In addition experiments have been reported which suggest that the transhydrogenase of Kaplan et al (8) may be operative in the ATP dependent reduction of TPN. The similar inhibition pattern observed with triiodothyronine as well as the similar temperature effect on the reaction support this conclusion.

REFERENCES

1. Chance, B. and Hollunger, G., J. Biol. Chem., 236, 1534 (1961).
2. Klingenberg, M. and Schollmeyer, P., Proc. of the Vth Internatl. Cong. of Biochem., Pergamon Press, 5, 1961, p. 46.
3. Estabrook, R.W., Fugmann, U., and Chance, E.M., Proc. of the Vth Internatl. Cong. of Biochem., Pergamon Press, (1961), p. 464.
4. Klingenberg, M., in 11th Mosbach Colloquium on Free Nucleotides and Biological Function, Mosbach, Springer-Verlag, April 1960, p. 82.
5. Estabrook, R.W., Anal. Biochem., 4, 231 (1962).
6. Estabrook, R.W. and Maitra, P.K., Anal. Biochem., 3, 369 (1962).
7. Danielson, L. and Ernster, L., Biochem. Biophys. Res. Comm., 10, 91 (1963).
8. Kaplan, N.O., Colowick, S.P., and Neufeld, E.F., J. Biol. Chem., 205, 1 (1953).
9. Ball, E.G. and Cooper, O., Proc. Natl. Acad. Sci. U.S., 43, 357 (1959).
10. Stein, A.M., Kaplan, N.O. and Ciotti, M.M., J. Biol. Chem. 234, 973 (1959).
11. Kaplan, N.O., personal communication.
12. Hommes, F. and Estabrook, R.W., Biochem. Biophys. Res. Comm., 11, 1 (1963).

DISCUSSION

Sanadi: Would you clarify your statement on the influence of magnesium on the transhydrogenase reaction?

Estabrook: I can describe the experiment, even though we do not know the full interpretation. These are Dr. Hommes' experiments. We were most interested in Dr. Danielson's experiments since Dr. Hommes had similar particles. We tried the system that Dr. Danielson described, and then investigated the effect of triiodothyronine (TIT) on the reaction. TIT proved to be inhibitory, and we considered the possibility of an antagonistic role between TIT and magnesium.

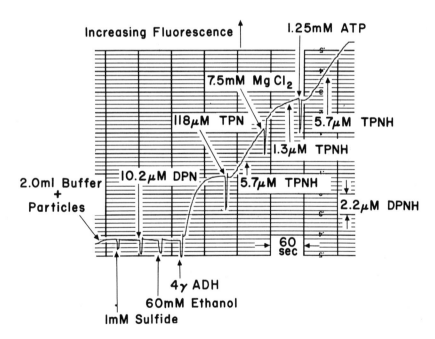

Fig. 1 (Estabrook). Since we had always carried out our experiments in the presence of magnesium, Hommes now performed the following experiment in the absence of magnesium. Particles were diluted in buffer and sulfide; alcohol, DPN, and alcohol dehydrogenase were added as described previously. The DPN becomes reduced. If TPN is added the reaction is initiated

without the addition of ATP. If we now add magnesium, the reduction of TPN stops. If we then add ATP, the role of TPN reduction is restored. The restoration of activity by ATP could not be replaced by versene.

Griffiths: Did you have ATP already there?

Estabrook: No. We visualized the reaction as follows: DPN is transformed to an activated DPN, whose binding is affected by magnesium and ATP. The activated DPN can then participate in the transhydrogenase reaction. In other words, my Fig. 7 may be modified to include an additional step where DPN or DPNH is converted to a bound DPN or DPNH, and it is here that ATP is involved in the transhydrogenase reaction.

Sanadi: Do you have serum albumin in the reaction mixture?

Estabrook: No.

Conover: Bovine serum albumin has no effect on the reduction of DPN in the absence of magnesium. It does, however, stimulate the reaction in the presence of magnesium and ATP. This may be correlated with the presence of low levels of uncoupling fatty acids.

Klingenberg: I have made an experiment which bears on the energy requirements for this reaction. From the redox potential, it would follow logically that this reaction would not require one ATP.

Estabrook: The redox potential difference between the DPN couple and the TPN couple is 50 mV.

Klingenberg: Yes, but the ATP provides about 300 mV. In Fig. 2 (which is also Fig. 4 of my paper) we show how the reduction of TPN is accelerated by ATP, and then is titrated back by phosphate and ADP. The hydrogen comes from succinate, which first has to reduce DPN by the "reversed reduction." If we then provide hydrogen from hydroxybutyrate, the extensive reduction of the TPN is obtained at a phosphorylation potential where the succinate-linked TPN reduction is no longer possible. There is also the interesting observation that the presence of rather low concentrations of uncoupler completely prevents reversal of oxidative phosphorylation, even though one can still induce the ATP-linked transhydrogenase reaction. This points out that the primary mechanism is probably the same but the quantitative result, of course, may be different. Also, the sensitivity to oligomycin is stronger for the succinate-linked TPN reduction which is completely inhibited by oligomycin, while the β-hydroxybutyrate-linked TPN reduction is not blocked by oligomycin, but only slowed down.

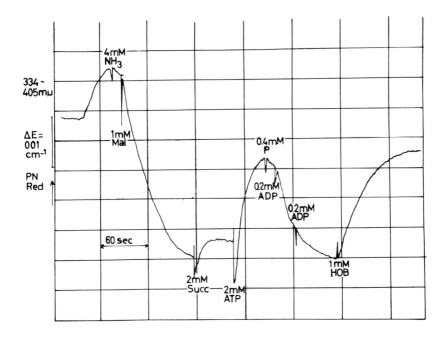

Fig. 2 (Klingenberg).

Estabrook: The subsequent ATP reduction of TPN of the type we showed was inhibited by oligomycin.

Klingenberg: The reduction goes slower with oligomycin, but it may reach a larger extent.

Estabrook: Dr. Klingenberg, do you feel that a phosphate potential is operative here in the same manner as it is in the succinate reduction? This is what we think, but we have no direct evidence in support of this.

Klingenberg: I think that, in fact, it is operating. But apparently a smaller phosphorylation potential is required to provide energy for TPN reduction. The latter reaction is a conventional measure of reversed energy transfer. However, the mechanisms of reversed energy transfer may be the same with both substrates. This may be explained if one considers that the extent of TPN reduction should depend on the product of the DPN reduction and on the phosphorylation potential. The reduction of DPN by succinate requires a higher phosphorylation potential than that of hydroxybutyrate, which can reduce DPN without energy supply.

Mildvan: In your last experiment you reported that large amounts of TPNH were produced. This rules out the trivial possibility (which might not be so easy to rule out in mitochondria) that there is a direct phosphorylation of DPN to produce TPN. Also, are you certain that the action of TIT is specific, because there are several dehydrogenases which are totally inhibited by thyroxine?

Estabrook: We cannot say it is specific. To my knowledge there are only two characteristics of the transhydrogenase which we can test; one is the inhibition by TIT and the other is the high activation energy for the reaction. Perhaps it is fortuitous in your terms that they come out to be exactly the same. When we do analyses of DPN, TPN, DPNH and TPNH we make sure that the balance of pyridine nucleotide recovery is correct. We have also checked our method with Dr. Pressman's technique and the results seem to correlate. We feel that we are getting within 10 per cent reproducibility in our results.

ENERGY-DEPENDENT REDUCTION OF TPN BY DPNH
Lennart Danielson and Lars Ernster
The Wenner-Gren Institute, University of Stockholm
Stockholm, Sweden

In 1959, Klingenberg and Slenczka (1) made the important observation that incubation of isolated liver mitochondria with DPN-specific substrates or succinate in the absence of phosphate acceptor resulted in a rapid and almost complete reduction of the intramitochondrial TPN. These and related findings led Klingenberg and co-workers (1-3) to postulate the occurrence of an ATP-controlled transhydrogenase reaction catalyzing the reduction of mitochondrial TPN by DPNH. A similar conclusion was reached by Estabrook and Nissley (4).

The present paper describes the demonstration and some properties of an energy-dependent reduction of TPN by DPNH, catalyzed by submitochondrial particles. Preliminary reports of some of these results have already appeared (5, 6), and a complete account is being published elsewhere (7).

We have studied the energy-dependent reduction of TPN by DPNH with submitochondrial particles from both rat liver and beef heart. Rat liver particles were prepared essentially according to the method of Kielley and Bronk (8), and beef heart particles by the method of Löw and Vallin (9). Below,

---

Abbreviations: AMP, ADP, ATP: adenosine 5'-mono, di-, triphosphate; CTP, GTP, ITP, UTP: cytidine, guanosine, inosine, and uridine 5'-triphosphate; DPN, DPNH, TPN, TPNH: di- and triphosphopyridine nucleotide, oxidized and reduced forms; P_i: inorganic orthophosphate; ADH: alcohol dehydrogenase; GSSG: oxidized glutathione.

This work has been supported by grants from the Swedish Cancer Society. We thank Miss Barbro Häggmark for excellent technical assistance.

only the data obtained with rat liver will be presented. The results with beef heart were essentially analogous (cf. 5).

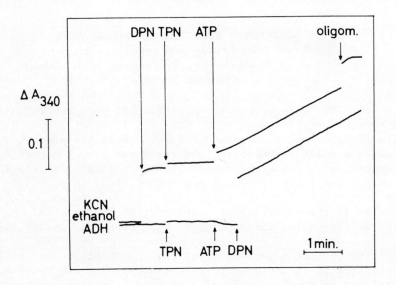

Fig. 1. ATP-dependent reduction of TPN by DPNH. The reaction mixture consisted of 50 mM tris-buffer pH 8.0, 6 mM $MgCl_2$, 250 mM sucrose, 1 mM KCN, 57 mM ethanol, 0.25 mg ADH, and particles containing 0.65 mg protein. Further additions were: DPN, 0.0167 mM; TPN, 0.2 mM; ATP, 2 mM; oligomycin, 3 µg. The final volume was 3 ml, and the temperature 30° C.

Rat liver particles were incubated in a buffered, Mg^{++} containing medium in the presence of KCN, ethanol, ADH, and 0.05 µmoles of DPN (Fig. 1). When the reduction of DPN by the alcohol dehydrogenase system was completed, 0.7 µmoles of TPN was added. No appreciable increase in A_{340} occurred. When 6 µmoles of ATP were now added, A_{340} increased at a linear rate. No similar increase in A_{340} was observed if DPN or TPN was omitted. Similar results were obtained when DPNH rather than DPN was added at the start, or when ethanol and ADH were replaced by β-hydroxybutyrate, inasmuch as the particles contained a sufficiently active β-hydroxybutyric dehydrogenase

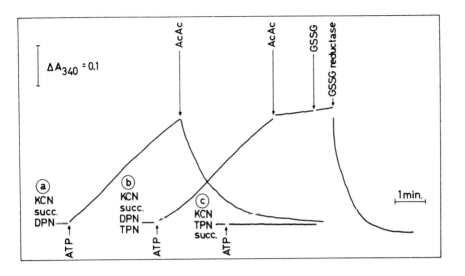

Fig. 2. ATP-dependent reduction of TPN by DPNH, coupled to the reduction of DPN by succinate. The reaction mixture consisted of 50 mM tris-buffer pH 8.0, 6 mM $MgCl_2$, 250 mM sucrose, 1 mM KCN, 3.3 mM succinate, and particles containing 0.63 mg protein. Further additions were: DPN, 0.167 mM; TPN, 0.2 mM; ATP, 2 mM; acetoacetate (AcAc), 5.7 mM; oxidized glutathione (GSSG), 3.3 mM; GSSG reductase in an amount capable of oxidizing 0.5 μmole TPNH/min. Final volume, 3 ml; temperature, 30° C.

The particles exhibited an ATP-dependent reduction of DPN by succinate (Fig. 2a). The formation of DPNH could be ascertained by adding acetoacetate which caused a rapid drop in A_{340}. When TPN was added together with DPN (Fig. 2b), the same rate of pyridine nucleotide reduction was observed as with DPN alone. However, in this case, addition of acetoacetate caused no drop in A_{340} but this occurred when GSSG and a purified preparation of TPNH-specific GSSG reductase were added. The reduction of TPN by succinate proceeded via DPN, as shown by the fact, that when DPN was omitted, no reduction of TPN occurred (Fig. 2c).

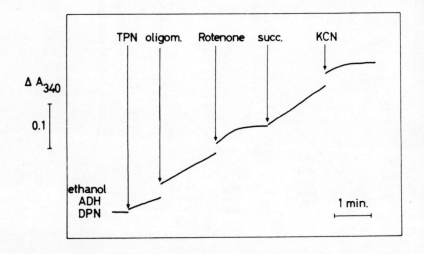

Fig. 3. DPNH-linked TPN reduction supported by aerobically generated high-energy intermediates. The reaction mixture consisted of 50 mM tris-buffer pH 8.0, 6 mM $MgCl_2$, 250 mM sucrose, 57 mM ethanol, 0.1 mg ADH, 0.033 mM DPN, and particles containing 0.65 mg protein. Further additions were: TPN, 0.233 mM; oligomycin, 3 µg; rotenone, 2 µg; succinate, 3.3 mM; KCN, 1 mM; ATP, 2 mM. Final volume, 3 ml. Temperature, 30°C.

Energy-dependent reduction of TPN by DPNH could also be achieved under aerobic conditions. In this system, addition of ATP was not required, the energy being supplied by the aerobic oxidation of DPNH (Fig. 3). When the latter was blocked, e.g. by rotenone, TPN reduction was abolished. Addition of succinate to the rotenone blocked system restored TPN reduction, which again could be abolished by further addition of cyanide. As could be anticipated the DPNH-linked TPN reduction was under all conditions insensitive to both rotenone and amytal; this was in contrast to the succinate-linked DPN reduction which was abolished by both agents. Both the DPNH-linked TPN reduction and the succinate-linked DPN reduction were insensitive to antimycin A.

Oligomycin did not inhibit the energy dependent TPN reduction under aerobic conditions; it even stimulated the reaction both when DPNH and succinate generated the high-energy intermediates. P_i and ADP had no effect on the reaction when added separately but depressed it somewhat when added in combination; this effect was abolished by oligomycin.

The energy-transfer system involved in the DPNH-linked TPN reduction supported by aerobically generated high-energy intermediates showed no requirement for P_i, as could be ascertained by using phosphate-free particles and showing that added P_i had no effect on the DPNH-linked TPN reduction. The reaction also was found to be remarkably insensitive to arsenate (5 mM) both in the absence and in the presence of oligomycin.

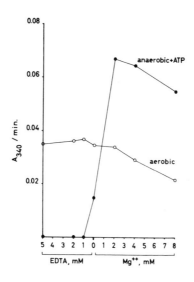

Fig. 4. Effects of Mg^{++} and EDTA on the energy-dependent reduction of TPN by DPNH. Conditions for the "anaerobic + ATP" system were as in Fig. 1, and for the "aerobic" system as in Fig. 3 (with succinate and rotenone but without oligomycin). In both cases DPNH rather than DPN was added. The amount of particle protein used per cuvette was 0.75 mg.

The ATP-supported, anaerobic, reduction of TPN by DPNH was greatly dependent on the concentration of Mg^{++} in the medium (Fig. 4). Maximal stimulation, about 4-fold over that of the system without added Mg^{++}, occurred at a Mg^{++} concentration of 2 mM. The slight activity of the system without added Mg^{++} could be abolished by 1 mM EDTA.

The DPNH-linked TPN reduction supported by aerobically generated high-energy intermediates was, in contrast to the ATP-supported system, independent of added Mg^{++}, and it was insensitive to EDTA up to a concentration of 5 mM (Fig. 4).

Table I

Comparison of nucleoside triphosphate-dependent TPN reduction by DPNH and nucleoside triphosphatase activities

Conditions were as in Fig. 1, except that DPNH rather than DPN was added. Each nucleoside triphosphate was added in a final concentration of 2 mM. Pyridine nucleotide reduction was followed for 5 min., after which the samples were fixed with perchloric acid for determination of P_i.

Nucleoside Triphosphate	TPN reduced μmoles/5min./ mg protein	P_i liberated μmoles/5min./ mg protein
ATP	0.331	3.30
ITP	0.141	1.66
GTP	0.140	1.42
CTP	0.047	0.19
UTP	0.044	0.36

ITP, GTP, UTP and CTP could replace ATP to varying extents in supporting DPNH-linked reduction (Table I). The relative activities obtained with the four nucleoside triphosphates in comparison to ATP were similar to the corresponding nucleoside triphosphatase activities, ITP and GTP being about half as active as ATP, and UTP and CTP considerably less active. The difference in activities between the different nucleoside triphosphates could not be overcome by increasing their concentrations, as illustrated in Fig. 5.

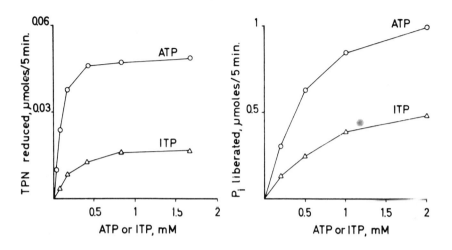

Fig. 5. <u>Energy-dependent reduction of TPN by DPNH and nucleoside triphosphatase activities with different concentrations of ATP or ITP</u>. Conditions were as in T$_a$ble I.

In Fig. 6 we compare the abilities of ATP and ITP to support DPNH-linked TPN reduction, succinate-linked DPN reduction and nucleoside triphosphatase activities, as well as to undergo exchange reactions with inorganic phosphate. The triphosphatase and exchange activities were measured in both the presence and absence of DPNH + TPN or succinate + DPN; no significant changes in the activities due to these additions were observed. The succinate-linked DPN reduction proceeded only at a low rate with ITP, or in some experiments even not at all (see also Löw and Vallin (9)). Likewise ITP exhibited no or only very slow phosphate exchange reaction. The relative ability of ITP to act in these two reactions thus was strikingly weaker than that found in the case of the DPNH-linked TPN reduction or the nucleoside triphosphatase.

Oligomycin inhibited the DPNH-linked TPN reduction supported by either ATP or ITP (or other nucleoside triphosphates) as well as the corresponding triphosphatase activities; it also inhibited the ATP-dependent succinate-linked DPN reduction, in accordance with previous reports. Comparison of the effects of increasing amounts of oligomycin on these reactions revealed no significant difference between their oligomycin sensitivities (Fig. 7). The DPNH-linked TPN reduction supported by aerobically generated high energy intermediates was, as already pointed out, insensitive to oligomycin.

Atractyloside, up to a concentration of 1 mM, did not inhibit any of the reactions studied.

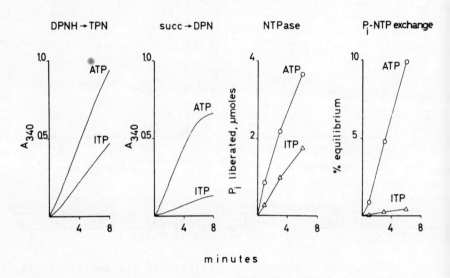

Fig. 6. Comparison of DPNH-linked TPN reduction, succinate-linked DPN reduction, nucleoside triphosphatase (NTPase), and P_i-NTP exchange activities with ATP and ITP. DPNH-linked TPN reduction was assayed essentially as described in Fig. 1, and succinate-linked DPN reduction as in Fig. 2a. ATP and ITP were added in final concentrations of 2 mM, and P^{32} in a final concentration of 0.16 mM. NTPase and P_i-NTP exchange were assayed both under the conditions of DPNH-linked TPN reduction and succinate-linked DPN reduction, as well as in the absence of pyridine nucleotides and succinate; the values given in the Figure are mean-values of the three types of assay.

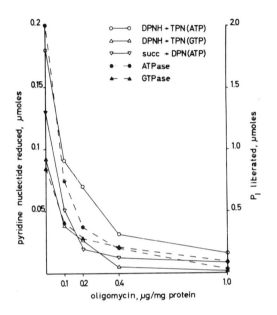

Fig. 7. Comparison of effects of oligomycin on nucleoside triphosphate-dependent reduction of TPN by DPNH and of DPN by succinate, as well as on nucleoside triphosphatase activities. DPNH-linked TPN reduction was assayed as in Fig. 1, succinate-linked DPN reduction as in Fig. 2a, and nucleoside triphosphatase as in Table I.

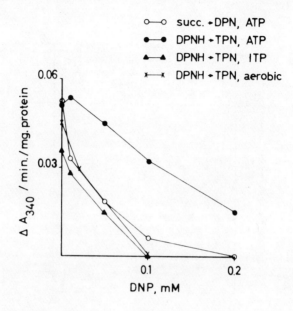

Fig. 8. Comparison of effects of 2,4-dinitrophenol (DNP) on DPNH-linked TPN reduction supported by ATP, ITP and by aerobically generated high-energy intermediates and of succinate-linked DPN reduction supported by ATP. ATP- or ITP- supported DPNH-linked TPN reduction were assayed as described in Fig. 1 and Table I, and that supported by aerobically generated high-energy intermediates as described in Fig. 3 (with succinate and rotenone but without oligomycin). ATP-supported succinate-linked DPN reduction was assayed as described in Fig. 2a.

The aerobic DPNH-linked TPN reduction was suppressed by 2,4-dinitrophenol to an extent similar to that usually found with respiratory chain-linked phosphorylations, 0.1 mM dinitrophenol giving full inhibition (Fig. 8). The same was true for the ATP-supported reduction of DPN by succinate (cf. also ref. 9) as well as for the ITP-supported reduction of TPN by DPNH. However, the DPNH-linked TPN reduction supported by ATP was markedly less sensitive to dinitrophenol, as shown in Fig. 8; 0.1 mM dinitrophenol inhibited this reaction less than 50 % (or in some experiments not at all), and 0.2 mM still allowed a significant reduction to take place.

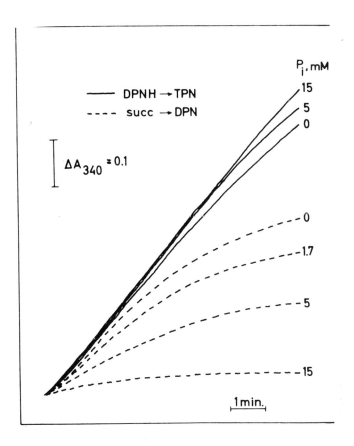

Fig. 9. <u>Effects of P_i on ATP-dependent reduction of TPN by DPNH and of DPN by succinate</u>. DPNH-linked TPN reduction was assayed as in Fig. 1, and succinate-linked DPN reduction as in Fig. 2a.

The ATP-dependent reduction of DPN by succinate declined sharply with time and the decline could be accentuated by the addition of P_i (Fig. 9). In contrast, the ATP-dependent reduction of TPN by DPNH proceeded at a virtually linear rate, which was unaffected by P_i.

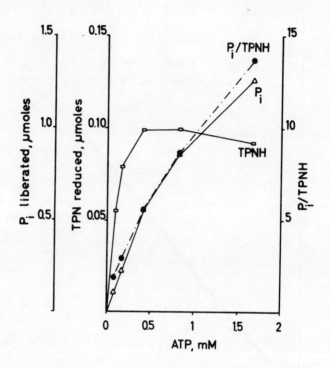

Fig. 10. **Estimation of P_i/TPNH ratio.** Particles were incubated in the presence of varying concentrations of ATP. Other conditions were as in Table I.

Under the standard conditions employed above, using 2 mM ATP, the ratio of μmoles P_i liberated to μmoles TPN reduced was of the order of 10-20.[1] This ratio could be diminished considerably by using lower concentrations of ATP as shown in Fig. 10. The lowest concentration of ATP that allowed accurate determinations of P_i and TPNH with the techniques used was 0.08 mM. With this concentration of ATP the P_i/TPNH ratios consistently were below 2; also consistently the values exceeded 1.

The particles catalyzed the reduction of DPN by TPNH. This could be demonstrated in different ways, two of which are illustrated in Table II. One system consisted of DPN, TPNH, glucose-6-phosphate and purified glucose-6-phosphate dehydro-

TABLE II

Comparison of requirements and capacities of the
submitochondrial particles to catalyze reduction
of TPN by DPNH and DPN by TPNH

DPNH-linked TPN reduction was assayed essentially as described
in Fig. 1. When indicated, DL-β-hydroxybutyrate (βOH) was
used in a final concentration of 6.7 mM. The assay system for
TPNH-linked DPN reduction consisted of 50 mM tris-buffer pH
8.0, 6 mM $MgCl_2$, 250 mM sucrose, 1 mM KCN, 0.2 mM (or as indicated) TPNH, 0.167 mM (or as indicated) DPN, and either 6.7 mM
glucose-6-phosphate (G-6-P) and an amount of glucose-6-phos-
phate dehydrogenase (G-6-PDH) capable of reducing 0.4 μmoles
TPN/min., or 5.7 mM acetoacetate (AcAc). When indicated, 2 mM
ATP was added. All assays contained 0.29 particle protein.

Reaction Studied	System	Pyridine nucleotide studied A_{340}/min./g protein	
		-ATP	+ATP
DPNH TPN	ethanol, ADH, DPNH, TPN	0	183
"	βOH, DPNH, TPN	0	167
TPNH DPN	G-6-P, G-6-PDH, TPNH, DPN	41	0
"	AcAc, TPNH, DPN	38	27
"	AcAc, TPNH*, DPN	46	32
"	AcAc, TPNH, DPN**	46	46

* 2 x TPNH, ** 2 x DPN

genase, and the reduction of DPN was measured. Another system
contained DPN, TPNH and acetoacetate, and the oxidation of
TPNH via DPN (by way of the β-hydroxybutyric dehydrogenase)
was followed. The data in Table II allow two main conclusions: 1) that the reduction of DPN by TPNH did not require
ATP for maximal activity; on the contrary ATP could even inhibit the reaction; and 2) that the rate of the TPNH-linked
DPN reduction was considerably lower than the rate of the ATP-
dependent DPNH-linked TPN reduction when the latter was measured under conditions of maximal activity. This situation
could not be altered by varying the concentrations of the pyridine nucleotides in the TPNH-linked DPN reduction system.
It may also be mentioned that the reduction of DPN by TPNH was
accompanied by no esterification of phosphate as could be ascertained by adding P_i^{32} and ADP to the system together with

hexokinase and glucose and following the appearance of P^{32} in glucose-6-phosphate.

Attempts were made to decide whether the same electron-transfer enzyme was involved in the TPNH-linked DPN reduction and the energy-dependent reduction of TPN by DPNH, or whether the two reactions involved separate catalysts. Thyroxine analogues, which have been shown to inhibit pyridine nucleotide transhydrogenase (10, 11), were found to suppress both reactions. However, the same compounds also inhibited ATPase (12) and the succinate-linked DPN reduction (13, 14) in accordance with earlier reports. In general no compound has yet been found to inhibit either one or both types of the pyridine nucleotide transhydrogenase reactions in a specific manner.

Our results reported above demonstrate the occurrence in mitochondria of an enzyme system catalyzing an energy-dependent reduction of TPN by DPNH, coupled to the energy-transfer system of the respiratory chain. Details of the proposed reaction mechanism are given in the scheme in Fig. 11.

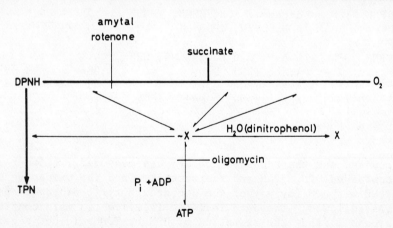

Fig. 11. Schematic representation of possible energy-transfer pathways involved in the energy-dependent reduction of TPN by DPNH catalyzed by submitochondrial particles.

The energy-dependent nature of the reaction and its coupling to the energy-transfer system of the respiratory chain is demonstrated by the requirement for ATP or for aerobically generated high-energy intermediates. The effects of oligomycin and dinitrophenol are also consistent with this conclusion

ENERGY-LINKED FUNCTIONS OF MITOCHONDRIA

The aerobic system is analogous to the system of succinate-linked acetoacetate reduction, previously described with intact mitochondria (15-21), in that the energy-transfer does not involve the participation of P_i and proceeds above the level of the oligomycin sensitive energy-transfer step. From the P_i/TPNH ratios reported in the results it appears that the energy requirement of the DPNH-linked TPN reduction probably is one high-energy bond equivalent per molecule of TPN reduced.

The energy-dependent reduction of TPN by DPNH differs from previously studied mitochondrial energy-linked reductions in that it does not represent the reversal of an oxidative phosphorylation. Phosphorylation connected with the oxidation of TPNH by DPN neither would be predicted to occur on thermodynamic grounds (22), nor is it known to occur experimentally. Hence, it is logical to expect that in the energy-dependent reduction of TPN by DPNH the major part of the invested energy is released as heat, and therefore, that the process is practically irreversible. This is in contrast to, e.g., the succinate-linked DPN reduction, where the bulk of the invested energy is conserved as a gain in oxidation-reduction potential. The data reported here on the difference in response to P_i and ADP between the two systems strongly support this conclusion.

The ability to proceed at appreciable rates with nucleoside triphosphates other than ATP, notably GTP and ITP, is an interesting feature of the energy-dependent DPNH-linked TPN reduction. Hitherto, only the Mg^{++} activated nucleoside triphosphatase reaction of submitochondrial particles has been known to possess this property (23). On the other hand, various partial reactions of oxidative phosphorylation, such as the P_i-ATP exchange reaction (24) and the succinate-linked DPN reduction (9), as well as oxidative phosphorylation itself (25, 26), have been shown to be strictly specific for adenine nucleotides. Our data are essentially in accordance with this picture although they indicate that the adenine nucleotide specificity, even in the case of the latter reactions, may be less strict than has been thought. Because of the nucleoside unspecificity of the Mg^{++}-activated triphosphatase it has been considered (23) that this enzyme may not be related to the energy-transfer system of respiratory chain-linked phosphorylations. This possibility can now be eliminated on the following grounds: 1. Oligomycin inhibits the reaction. 2. The DPNH-linked TPN reduction, which shows the same nucleoside unspecificity as the nucleoside triphosphatase, must be assumed to be related to the respiratory chain-linked energy-transfer system, since it can also proceed directly at the expense of aerobically generated high-energy intermediates.

Our tentative explanation for the fact that the DPNH-linked TPN reduction follows more closely the specificity pattern of the nucleoside triphosphatase than does the succinate-linked DPN reduction is based on the following kinetic reasoning: Since the nucleoside triphosphatase exhibits a greater maximal velocity with ATP than with other nucleoside triphosphates it follows that the steady-state level of $\sim X$ (see Fig. 11) will be lower with these than with ATP. The rate of the P_i-exchange reaction, as measured under the conditions here employed, ought to be a function of the steady state-level of $\sim X$, and therefore, the virtual lack of exchange in the case of e. g., ITP, is consistent with the assumption that the steady-state level of $\sim X$ in this system indeed must have been very low. From the experimental data it appears, furthermore, that the DPNH-linked TPN reduction possesses a particularly low requirement for $\sim X$, as shown by the P_i/TPNH data, as well as by its relative insensitivity to dinitrophenol when ATP--but not ITP--was used as the source of energy. The succinate-linked reduction of DPN, on the other hand, is highly sensitive to dinitrophenol and to P_i and ADP, indicating that its requirement for $\sim X$ is relatively high. Hence it is understandable that the DPNH-linked TPN reduction with its low requirement for $\sim X$, in terms of steady-state level, can proceed relatively better with ITP than can the succinate-linked DPN reduction with its high requirement.

Little can thus far be settled about the nature of the electron-transfer catalyst involved in the energy-dependent reduction of TPN by DPNH and its relationship to the mitochondrial pyridine nucleotide transhydrogenase studied by Kaplan et al. (27). Whether the electron transfer catalysts involved in the two reactions are the same or different cannot, however, be decided with the information available. Another point of interest is the nature of the intermediate(?) involved in the reaction. Logically, there ought to exist an intermediate between $\sim X$ and either the electron transfer catalyst itself or one of the pyridine nucleotides. Indications for the existence of high-energy derivatives of pyridine nucleotides have been reported in connection with studies of the mechanism of respiratory chain-linked phosphorylations (4, 28-33). Only one of these has so far been specified chemically, a compound postulated by Griffiths and Chaplain (32, 33) to consist of a phosphorylated form of DPNH. However, the fact that the DPN-linked TPN reduction in our aerobic system could proceed without the participation of P_i makes it improbable that DPNH-phosphate is an intermediate in our reaction.

In previous studies of the succinate-linked reduction of acetoacetate it was first shown (16, 17) that non-phosphorylated high-energy intermediates generated by the respiratory chain could be directly utilized for energy-requiring reactions in the mitochondria. The present reaction is the second example of such an energy-utilizing process, and the first of an energy-dependent reduction which is not the mere reversal of an energy-coupling oxidation.

REFERENCES

1. Klingenberg, M., and Slenczka, W., Biochem. Z., <u>331</u>, 486 (1959).

2. Klingenberg, M., II. Colloquium der Gesellschaft für physiologische Chemie, Mosbach 1960, Springer/Berlin-Göttingen-Heidelberg, p. 82.

3. Klingenberg, M., and Schollmeyer, P., in Symposium on Intracellular Respiration, Vth Internatl. Congr. Biochem., Moscow 1961, Pergamon, Oxford 1963, Vol. V, p. 46.

4. Estabrook, R. W., and Nissley, S. P., in Symposium über die Funktionelle und Morphologische Organisation der Zelle, Rottach-Egern 1962, Springer/Berlin-Göttingen-Heidelberg, 1963, p. 119.

5. Danielson, L., and Ernster, L., Biochem. Biophys. Res. Comm., <u>10</u>, 91 (1963).

6. Danielson, L., and Ernster, L., Proc. Scand. Biochem. Soc. Meeting, Copenhagen 1963, Acta Chem. Scand., in press.

7. Danielson, L., and Ernster, L., Biochem. Z., in press.

8. Kielley, W. W., and Bronk, J. R., J. Biol. Chem., <u>230</u>, 521 (1958).

9. Löw, H., and Vallin, I., Biochim. Biophys. Acta <u>69</u>, 361 (1963).

10. Ball, E. G., and Cooper, O., Proc. Natl. Acad. Sci. U. S., <u>10</u>, 91 (1959).

11. Stein, A. M., Kaplan, N. O., and Ciotti, M. M., J. Biol. Chem. <u>234</u>, 973 (1959).

12. Lindberg, O., Löw, H., Conover, T. E., and Ernster, L., in Symposium on Biological Structure and Function, Stockholm 1960, Academic Press, London 1961, Vol. II, p. 3.

13. Chance, B., and Hollunger, G., Nature, 185, 666 (1960).
14. Chance, B., and Hollunger, G., J. Biol. Chem., 238, 445 (1963).
15. Ernster, L., in Symposium on Biological Structure and Function, Stockholm 1960, Academic Press, London 1961, Vol. II, p. 139.
16. Ernster, L., in Symposium on Intracellular Respiration, Vth Internatl. Congr. Biochem., Moscow 1961, Pergamon, Oxford 1963, Vol. V, p. 115.
17. Ernster, L., in Symposium über die Funktionelle und Morphologische Organisation der Zelle, Rottach-Egern 1962, Springer/Berlin-Göttingen-Heidelberg, 1963, p. 98.
18. Azzone, G. F., Ernster, L., and Weinbach, E. C., J. Biol. Chem. 238, in press.
19. Ernster, L., Azzone, G. F., Danielson, L., and Weinbach, E. C., J. Biol. Chem., 238, in press.
20. Ernster, L., Azzone, G. F., Danielson, L., and Nordenbrand, K., Acta Chem. Scand., in press.
21. Klingenberg, M., and v. Häfen, H., Biochem. Z., 337, 120 (1963).
22. Rodkey, F. L., J. Biol. Chem., 234, 268 (1959).
23. Cooper, C., and Lehninger, A. L., J. Biol. Chem., 224, 54 (1957).
24. Cooper, C., and Lehninger, A. L., J. Biol. Chem., 224, 561 (1957).
25. Cooper, C., and Lehninger, A. L., J. Biol. Chem., 219, 48 519 (1956).
26. Devlin, T. M., and Lehninger, A. L., J. Biol. Chem., 219, 507 (1956).
27. Kaplan, N. O., Colowick, S. P., and Neufeld, E. F., J. Biol. Chem., 205, 1 (1953).
28. Purvis, J. L., Nature, 182, 711 (1958).
29. Chance, B., and Baltscheffsky, H., J. Biol. Chem., 233, 736 (1959).
30. Purvis, J. L., Biochim. Biophys. Acta, 38, 435 (1960).
31. Pinchot, G. B., and Hormanski, M., Proc. Natl. Acad. Sci. U. S. 48, 1970 (1962).

32. Griffiths, D.E., and Chaplain, R.A., Biochem. Biophys. Res. Comm., $\underline{8}$.

33. Griffiths, D.E. and Chaplain, R.A., Biochem. J., $\underline{85}$, 20P (1962).

DISCUSSION

<u>Slater</u>: A very interesting paper, indeed. When you say "a high or a low $\sim X$ experiment," do you mean a high or a low number of $\sim X$ molecules, or a high or a low energy of hydrolysis of $\sim X$?

<u>Danielson</u>: The requirement I was talking about is the requirement for the steady state level. When you use ATP, the steady state level of $\sim X$ is supposed to be higher than when you use ITP. The DPNH reduction has a low requirement for $\sim X$, as we postulate it on the basis of the P_i:TPNH ratio experiment. There you can see (cf. Fig. 10) that, if you diminish the ATP concentration, this reaction can compete with the other useless hydrolysis of $\sim X$ (i.e., ATP-ase).

<u>Slater</u>: My second point concerns the value of the ΔF or ΔG of the phosphate potential, using Klingenberg's terminology. The phosphate potential requirement is less for your pyridine nucleotide transhydrogenase than with succinate and DPN.

<u>Danielson</u>: Yes.

<u>Slater</u>: Then the stoichiometry is also a very important question. Are you sure that you really need a stoichiometric requirement of energy? What was the actual experimental evidence that you need one $\sim$ per molecule of pyridine nucleotide reduced?

<u>Danielson</u>: That is what is shown in Fig. 10.

<u>Slater</u>: It was bearing down towards zero very fast.

<u>Danielson</u>: Well, zero, no! We have shown that it is an energy-requiring process.

<u>Slater</u>: Well yes, but how near zero does it go? Is it a catalytic requirement? I would suggest that you have a catalytic requirement for this reaction, and that the stoichiometry is nearly zero.

<u>Danielson</u>: We just can't get less than one.

<u>Racker</u>: I think I have one comment which may shed some light on the findings of both Dr. Danielson and Dr. Estabrook. Depending on the type of submitochondrial particles, I found that I could duplicate both their findings. When I used

A-particles (obtained by sonic oscillation of mitochondria in 0.02 N ammonia) the ATP-linked transhydrogenation from DPNH to TPNH was dependent on F_1 as well as on F_4. The energy-dependent transhydrogenase in this case was only about one-fourth of the total activity. On the other hand, T-particles behave just like the preparations Dr. Estabrook has shown us; i.e., they show considerable transhydrogenase activity in the absence of both ATP and magnesium. Therefore, I think that there is room for both systems, and possibly this answers Dr. Slater's question also. Perhaps the transhydrogenase which depends upon a catalytic amount of ATP is the one inhibited by magnesium.

Klingenberg: I have always classified this reaction as ATP-controlled; it is neither ATP-dependent nor energy-dependent, but rather depends on the phosphorylation potential alone. In the same experiments under anaerobic conditions we can obtain TPN reduction without ATP. Under these conditions, if one avoids too extensive oxidation of the pyridine nucleotides, one can have a rather good reduction of the TPN again when β-hydroxybutyrate is added. (I speak here of the internal pyridine nucleotide of liver mitochondria). Although you may say that there is a residual amount of energy present, the pyridine nucleotide reduction comes to a level which should reverse down again.

Racker: May I point out that the ratio between the DPN and the TPN in my experiments was 1 to 100, and the DPN is kept reduced with ethanol and alcohol dehydrogenase. Therefore, there is very little oxidized DPN in our system, and this is an ATP- and coupling factor-dependent transhydrogenation.

Chance: What was the concentration of DPN to TPN?

Racker: The TPN concentration is 1 mM and the DPN concentration is 0.01 mM.

Klingenberg: In the particle system the DPN may be in a state comparable to that which exists in the mitochondria in a highly oxidized state.

Estabrook: You say your particle is very low in transhydrogenase. What do you use as a criterion for transhydrogenase?

Racker: It is the reaction in the absence of magnesium and ATP.

Estabrook: This is going from TPNH to DPN?

Racker: No, this is the same system that you as well as Dr. Danielson have been using, i.e., ethanol, alcohol dehydrogenase with catalytic amounts of DPN, and TPN as the hydrogen acceptor.

Estabrook: Then it is inhibited by magnesium?

Racker: In these A-particles the transhydrogenase activity in the absence of magnesium is very low. I mentioned that in the other particles it is quite active, and then I had trouble in demonstrating the coupling factor-dependent reaction.

Danielson: I may add that we never succeed in getting complete anaerobiosis by cyanide plus rotenone, so that we always get small reduction of TPN in these systems without having added any ATP. We think the reason for this is that the TPN is reduced from the generation of high-energy intermediates.

Racker: This is ruled out in our preparations, which work well without ATP even when oxidations are permitted to take place. We have trouble, moreover, with TPN reduction from the malate which arises from the oxidation of succinate. This is a possibility which has bothered us, and which may have occurred also in your studies. Succinate may be oxidized to malate, and the malate reduces TPN directly, or indirectly via DPN. This is presently occurring in some of our experiments and may even be energy-dependent, due to the need for removal of oxal acetate. In the presence of oligomycin the rate of his reaction is increased, which would be difficult to explain by a reaction going through malate.

Chance: Therefore, the reaction is definitely energy-dependent.

Racker: In Danielson's oxidative experiment, with succinate as substrate, he observed the reduction of TPN. Now my question is: how do we know that succinate doesn't go to malate, which serves as hydrogen donor?

Conover: It doesn't explain the TPNH.

Estabrook: The question is, where does the DPNH come from to reduce the TPN? Is it a succinate-linked reduction of DPN, or from malic dehydrogenase?

Racker: Yes. In fact, the reaction with malate may go via DPNH and be catalyzed by malic dehydrogenase, or go directly to TPN and be catalyzed by malic enzyme. We have run into both reactions in our systems.

Danielson: If you are referring to the aerobic system with rotenone and succinate, then in that case we had alcohol and alcohol dehydrogenase present all the time.

Chance: Yes, but I think Dr. Racker was just asking about the conditions where you did not have alcohol and alcohol dehydrogenase.

ENERGY-LINKED FUNCTIONS OF MITOCHONDRIA

Estabrook: Maybe he has given the answer: the succinate-linked reduction of DPN is rotenone-sensitive. Is your aerobic system rotenone-sensitive when succinate is present?

Danielson: No. We used DPNH as the hydrogen source; then we blocked its oxidation with rotenone, then we added succinate to generate the high-energy intermediate needed for the reaction.

Cohn: There has been confusion about the term "energy-linked." Some of you are using it in one sense and some in another. As Dr. Slater suggests, the stoichiometry is most important, and the real question is related to the reduction of DPNH and TPN. Is 1 molecule of ATP cleaved per H^+ transferred? What is the stoichiometry of ATP to ADP and P_i in the formation of TPNH? Do you observe the breakdown of 1 ATP? I haven't been able to figure out the answer to these questions from any of these comments.

Danielson: Maybe we have not shown this in a convincing way but this is what our data suggest.

Cohn: Is ADP + P_i formed in a ratio of one for every TPNH reduced?

Danielson: When you decrease the ATP concentration, yes.

Slater: (referring to Fig. 10) May I just comment on the determination of the stoichiometry in this experiment, where the amount of inorganic phosphate formed from ATP is related to the measurement of TPNH formed. Now the experimental data here show that the amount of ATP required at high concentrations is enormous; it is 10. There are many other ATP-ase reactions going on as well, and we don't know which ATP-ase reactions are operative.

Cohn: But is the ratio never less than 1?

Slater: No, he said not, but I said he was going down awfully close towards 1.

Danielson: We believe that the ATP requirement--even if you have 20 μmoles of ATP available per μmole of TPN to be reduced --is still 1.

Slater: Yes. When you have high concentrations of ATP you say it's 1 for your reaction and 19 in side reactions. When you have 1 μmole of ATP then you say it's 1 for your reaction and 0 for side reactions.

Danielson: Of course.

Slater: I suspect that you still have side reactions.

Racker: Dr. Cohn, we cannot answer your question in regard to our system of transhydrogenation because there is considerable ATP-ase activity present.

Cohn: Yes, but your evidence that you need a soluble factor, or that oligomycin sensitivity exists in the transhydrogenase reaction merely indicates that there is a common intermediate with other reactions of the system. It doesn't tell us anything about the energy requirements for the transhydrogenase reaction.

Racker: Yes, I fully agree.

Klingenberg: If we accept that glutamate dehydrogenase is reducing ketoglutarate plus NH_3 by TPNH, then it involves a transhydrogenase reaction going from malate to glutamate. Under anaerobic conditions one can have quite extensive glutamate formation from malate, if one only takes measures that the pyridine nucleotides do not get too highly oxidized for example by adding glutamate (cf. Fig. 2 of my paper). In this case one would have a high turnover of transhydrogenase without the addition of ATP.

Chance: Under anaerobic conditions and in the absence of an energy source.

Klingenberg: Yes.

Slater: Just confirming this point again: on the same assumption that our aspartate synthesis involves the transhydrogenase reaction, I have shown that we obtained a stoichiometry of 6 molecules of amino acid formed per one $\sim$ expended. This was the basis of the conclusion that we have only a catalytic requirement. If we can assume with Klingenberg that this involves the pyridine nucleotide transhydrogenase reaction, that is all I'm interested in.

Estabrook: You have no evidence to assume this.

Chance: And no evidence to refute it, either.

Slater: There is no evidence, that's true, but I just want to know how good his evidence is that it's otherwise.

Chance: Perhaps an interesting experiment would be to try TIT on the anaerobic system.

SPECIFIC INHIBITORS OF ENERGY TRANSFER
Berton C. Pressman
Johnson Research Foundation, University of Pennsylvania
Philadelphia, Pennsylvania

The overall mechanism by which energy arising from the electron transport chain is utilized for ATP synthesis, or some of the alternate energy consuming processes under discussion today, has actively concerned many research groups. One approach taken has been the derivation of particles from mitochondria which retain some partial function of the original complex, followed by an attempt to reconstruct the whole from the sum of the parts. Another source of jig-saw pieces from which to reconstruct the master plan is the isolation and identification of hitherto unknown components of the scheme. A third approach, which has had particular appeal to us, is the characterization of the energy transfer pathway complex from its interaction with a series of sequential site-specific agents. A propos of the general theme of this colloquium it will be seen that certain of our observations bear rather directly on the utilization of energy derived from the electron transport chain prior to the involvement of inorganic phosphate.

Lardy has already provided considerable insight into the mechanism of two types of energy transfer inhibitors:

1. Uncoupling agents, originally emphasizing DNP, later encompassing an extensive group of additional compounds,

which permit the dissipation, as heat, of energy otherwise obligatorily utilized for endogonic biological processes (1);

2. Oligomycin which blocks the availability of such energy for ATP synthesis (2). It was also reported by Hollunger (3) that guanidine blocked ATP synthesis in a manner resembling the action of oligomycin.

Our initial objective was to obtain a guanidine derivative more potent than the parent compound, which has the relatively poor Km for mitochondria of 5 mM. To this end we synthesized a series of guanidine derivatives and finally standardized on octylguanidine, which is 500 times as active as guanidine itself (Km, 10 μM) (4).

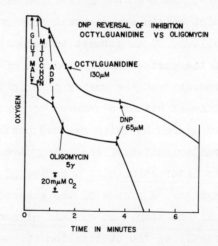

Fig. 1. Composite of oxygen electrode tracings depicting the onset of respiratory inhibitions by octylguanidine and oligomycin and their subsequent release by DNP; materials and methods as described previously by Pressman (4).

In Fig. 1 we call attention to some of the similarities and differences between the response of a state 3 rat liver mitochondrial system to octylguanidine on the one hand and oligomycin on the other. The inhibition induced by octylguanidine is relatively slow, and the speed of onset varies with the inhibitor concentration. On addition of an uncoupling agent such as DNP, following a time lag, there is a slow progressive restoration of respiration which often falls short of the original rate. Oligomycin induces rapid and nearly complete block of respiration, although some time delay can be detected, and upon addition of an uncoupling agent such as dinitrophenol or dicumarol there follows an instantaneous and complete release of the respiratory block. The short lag in response to oligomycin appears to be some type of permeability phenomenon peculiar to this compound since another antibiotic described by Lardy, aurovertin (5), seems to act like instant oligomycin. Under the same conditions, the onset of the inhibition by this agent appears instantaneously within the limits of the response time of our oxygen electrode.

In Fig. 2 we see that the same octylguanidine inhibited system fails to respond to the uncoupling agent dicumarol at a level which is fully as active as DNP in stimulating the respiration of a phosphate acceptor limited system (i.e., state 4). It thus appears that the octylguanidine inhibition unmasks a difference in action between the two uncoupling agents, dinitrophenol and dicumarol.

Hollunger had shown that guanidine exhibits a substrate specificity (3) which cannot be obtained with oligomycin. Guanidine and most of its derivatives are not nearly as

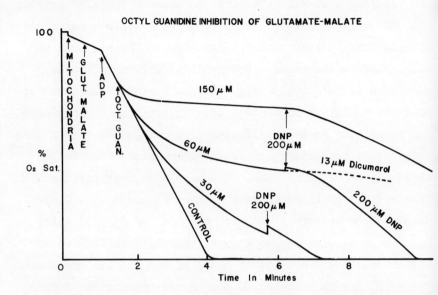

Fig. 2. Composite of tracings depicting the titration of mitochondria with octylguanidine and subsequent release by uncoupling agents.

effective in inhibiting succinate respiration as they are in inhibiting DPN-linked respiration, but high levels of guanidines do induce a partial inhibition of succinate respiration. In Fig. 3 we see that when succinate respiration is inhibited by guanidines, however, the inhibition is fully reversable by dicumarol as well as DNP. This establishes that the failure of dicumarol to act on the octylguanidine-inhibited DPN-linked substrate system is not due to any basic incompatibility between octylguanidine and dicumarol, but is rather indicative of the inherent loci of action of octylguanidine and the uncoupling agents.

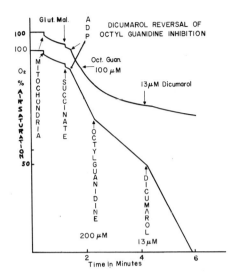

Fig. 3. Influence of substrate on the dicumarol reversal of octylguanidine inhibition.

We surveyed a large number of guanidine derivatives and found a small number which possessed a totally different substrate specificity from the majority, the best known of which is the oral hypoglycemic agent DBI (phenethyl biguanide). These compounds are actually *more* effective in inhibiting succinate respiration than they are in inhibiting DPN-linked substrate respiration. At first we concluded that this group acted specifically against succinate systems (6) until further data indicated that the situation was otherwise. The DPN-linked substrates we originally examined, β-hydroxybutyrate or glutamate, are apparently not capable of forming energized intermediates as effectively as does succinate. When we employed the DPN-linked glutamate-malate substrate systems, which can phosphorylate ADP at the

ENERGY-LINKED FUNCTIONS OF MITOCHONDRIA

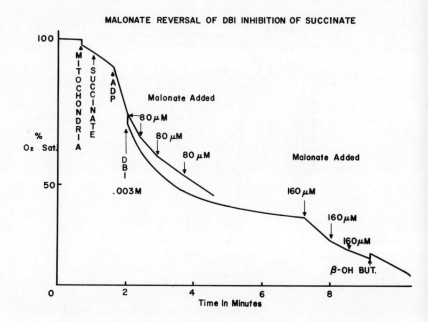

Fig. 4. Malonate reversal of DBI inhibition of succinate.

same rate as does succinate, we found that DBI inhibited respiration quite effectively. From this we deduce that DBI inhibits utilization of both DPN-linked substrates and succinate at a common point. Fig. 4 illustrates the reason that DBI is so poorly effective against our original DPN-linked substrates, namely the higher the respiratory rate obtained in the absence of the inhibitor, the stronger is the inhibition that sets in after the addition of the inhibitor. Here we see that upon addition of small amounts of malonate to a DBI-inhibited succinate system the inhibition is partially reversed. In other words limiting the potential rate at which succinate could push electrons into the electron transport chain actually permits more electrons to flow through the chain! Thus, if the substrate combination glutamate-

RESPONSE OF CARRIERS TO GUANIDINE INHIBITION

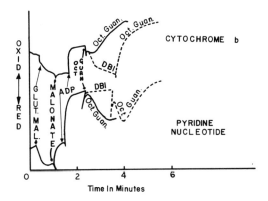

Fig. 5. Response of cytochrome b and pyridine nucleotides to DBI and octylguanidine response as revealed by double-beam fluorescence apparatus (7). In alternate experiments, at the point marked "octylguanidine," DBI was added instead and the resulting response represented by the dashed tracing superimposed over the octylguanidine response tracing. Subsequently octylguanidine was added to the DBI systems producing qualitatively the same effect as in the absence of DBI.

---

malate is used to achieve a DPN-linked, high electron generating capacity, the resulting system is inhibited by DBI as is the succinate system. Conversely, if malonate is used to reduce the electron generating capacity of succinate to that of β-hydroxybutyrate, the resulting system responds poorly to DBI. The differences between the action of these two classes of guanidine derivatives described are brought out even more dramatically by means of the double beam spectrophotometer-fluorescence apparatus of Chance (7). Rather than analyze Fig. 5 in detail, I would like to call your attention to the upper trace indicating the status of cytochrome b. Addition of octylguanidine produces an oxidation of cytochrome b indicative of a block on the <u>substrate</u> side of cytochrome b as reported previously by Chance and Hollunger (8). DBI addition, on the other hand (dashed tracing), leads to a reduction of cytochrome b, consistent with a block on the <u>oxygen</u> side of cytochrome b.

These experiments not only permit the assignment of different loci of action to the two types of energy transfer inhibiting guanidines, but also reveal that the uncoupling agents as well have site-specificity. Table I demonstrates that the specificity of dicumarol correlates only with the location of the energy transfer block rather than with the presence of particular substrates or inhibitors. According to the evidence already presented, DBI inhibits preferentially at the cytochrome b energy conservation site, regardless of the substrate employed, and we observe that the respiratory block is released by dicumarol in the case of either substrate. The succinate pathway does not involve the DPN-flavin energy conservation site and accordingly, regardless of the inhibitor employed, the respiratory block is released by dicumarol. Only in the case of the unequivocal DPN-flavin energy conservation site block, the octylguanidine-glutamate-malate system, do we obtain conditions appropriate for the demonstration of uncoupler specificity. In this instance dicumarol fails to release the respiratory inhibition, while DNP does.

We thought it appropriate to exploit this technique for unmasking the specificity of uncoupling action to survey a group of agents, of widely diverse structures, for their

TABLE I
Release of Energy Transfer Blocked Respiration by Dicumarol

Substrate	Inhibitor	
	Octylguanidine	DBI
Glutamate-Malate	NOT Released	Released
Succinate	Released	Released

Table I. Summary of influence of inhibitors and substrates on stimulation of respiration by dicumarol.

apparent site-specificity. We tested the ability of each uncoupling agent to act on both the DPN-flavin conservation site energy transfer pathway, i.e., respiratory stimulation of the glutamate-malate-octylguanidine system, and the cytochrome b conservation site transfer pathway, i.e., stimulation of the succinate-DBI system.

The first category, represented in Table II by DNP and pentachlorophenol, restores about 50 per cent of the uninhibited rate of the octylguanidine system and 100 per cent of

TABLE II

RELEASE OF ENERGY TRANSFER BLOCKS
BY UNCOUPLING AGENTS

Uncoupler	Relative Activity	% Release Oct. Glu.	% Release DBI
DNP	1	36	106
Penta-Cl-Phenol	2.1	54	86
m-Cl-CCP	160	24	107
CCP	41	12.5	112
Dicumarol	8	3.2	140
Octyl-DNP	6	2.2	130
Gramacidin	42	16	35(slow!)
Valinomycin	1200	19	19(slow!)

Comparison of the abilities of various uncouplers to release octylguanidine and DBI inhibited systems. Each uncoupler was employed at a concentration five times that necessary to stimulate an ADP deficient system to one-half maximal respiration. The relative activities of the uncouplers was compared on a similar basis, relative to DNP, which gave one-half maximal stimulation at 15 μM.

the uninhibited rate of the DBI system. Dinitro-o-cresol, as well as other nitrated and halogenated phenols, fall into this category. The second category, which might just as well have been considered a subcategory of the first, includes the carbonyldicyanophenylhydrazones (9) which are somewhat less effective in the octylguanidine system than are the first category, but, like the latter, are very effective in the DBI system. The third group is characterized by a very poor ability to reverse glutamate-malate-octylguanidine inhibition, but maintains full effectiveness in the succinate-DBI system. In addition to dicumarol already discussed, this category includes octyldinitrophenol (10) and the lichen-derived complex phenol, usnic acid (11).

We were most surprised to find yet another group of uncouplers, the existence of which was unexpected prior to the survey. This last group consists of gramicidin (1) and valinomycin (12), which are respectively cyclic polypeptide or depsipeptide (13) compounds. This group is extremely effective in stimulating the respiration of phosphate acceptor limited systems, but is very poor in releasing either type of guanidine derivative inhibition. The respiratory release obtained in either case is not only incomplete, but occurs only after a considerable time delay. If anything, these inhibitors appear more effective in the octylguanidine system than in the DBI system.

Fig. 6 presents our attempt to account for the behavior of the various types of inhibitors and uncouplers in terms of specific loci of action along the energy transfer pathway leading to the synthesis of ATP. I prefer not to go into the details of our rationale at this time (it rests heavily on the

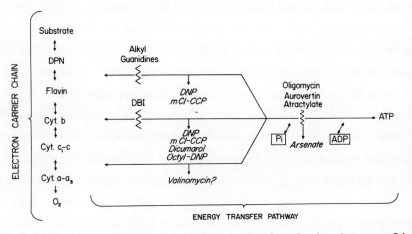

Fig. 6. General scheme indicating relative lock of uncoupling agents and inhibitors along the transfer pathway of energy arising from the electron carrier chain.

---

nature of ATP-ase induction in the guanidine blocked systems and has been presented in detail in a recent publication), but we conclude that, despite superficial similarities between the action of the guanidines and oligomycin, their respective points of action occur on opposite sides of the locus of action of the uncouplers. Aurovertin (5) and atractylate (14) have been reported to share many properties with oligomycin, and have accordingly been placed at the same locus. This scheme accounts for the substrate-specificity of the guanidine derivatives. Their locus of action occurs **before** the junction of the energy transfer pathways arising from the separate energy conservation sites to form a common pathway for the synthesis of ATP. Since oligomycin acts only after the junction, its action would not be expected to reflect any specificity related to the site-specific origin of energy. We infer that uncouplers, such as DNP and the carbonyldicyanophenylhydrazones, affect analogous loci on both of the upper two branches, but dicumarol and octyldinitrophenol do not act on the upper branch nearly as effectively as they do on the middle branch. In order to account for the unique properties of the last uncoupler category, we have tentatively placed it along the bottom branch. Valinomycin obviously reacts instantaneously with some component of the energy transfer pathway, since it evokes an instantaneous stimulation of phosphate-acceptor limited systems, but, by the criteria we have set up, this hypothetical component is not on either the upper or middle energy transfer branch. For technical reasons we are not in a position to test directly whether or not an uncoupler acts along the lower energy transfer branch. To attempt to do this we turned to the ascorbate-TMPD system of Dr. Packer (15) to feed energy obligatorily into the bottom branch. Although we obtained virtually the same degree of respiratory control and oligomycin sensitivity as reported by Dr. Packer, we were unsuccessful in obtaining a block with any of the site-specific guanidines, which would have been a prerequisite for setting up an appropriate test system.

We would like to acknowledge an aspect of our scheme which cannot be completely rationalized as yet. We do not know for certain the mechanism by which the specificity of the guanidines on the substrate side of the uncoupling agent is in turn conferred on the locus of action of the latter, but nevertheless this is what we observe.

The differences in uncoupling action between valinomycin and uncouplers of the other groups are fundamental enough to be manifested even in the absence of the site-specific

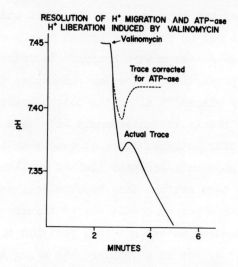

Fig. 7. Recording of pH response to valinomycin of a mitochondrial system (ATP, $MgCl_2$, KCl, tris, sucrose). The solid trace was obtained experimentally. The dashed curve was derived from this by correcting for H^+ production due to ATPase, as determined directly by phosphate analysis.

---

energy transfer blocks. We received quite a surprise when we began to examine the stimulation of mitochondrial ATP-ase by valinomycin. We followed ATP hydrolysis by means of the recording pH meter and obtained the rather bizarre drain pipe shaped tracing shown in Fig. 7. Upon addition of valinomycin we obtained a rapid fall in pH followed by a reversal of pH change which contributes the elbow of the drain, and finally the pH begins to fall again at a rate slower than the initial rate. When the splitting of ATP was measured directly by

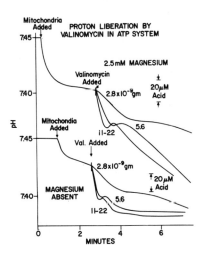

Fig. 8. Influence of valinomycin concentration ± magnesium on the rate and extent of H^+ ejection from mitochondria.

---

phosphate analysis we found it occurred at a constant rate during all phases of the pH tracing. The change in pH during the last segment of the curve was fully accountable for on the basis of the known acid liberation accompanying the splitting of ATP. Since we could ascertain that ATP hydrolysis took place at a constant rate, we subtracted this acid production from the original tracing and derived the dashed curve which represents a triphasic migration of H^+ in and out of the mitochondria which occurs concomitantly with the hydrolysis of ATP. Fig. 8 shows that, although the rate of this initial H^+ expulsion from the mitochondria is a function of the amount of valinomycin added, the final extent of the process is not and is determined rather by some inherent

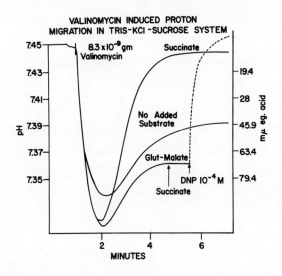

Fig. 9. Direct recording of valinomycin-triggered H^+ ejection in Tris (.02 M)-KCl-sucrose medium.

---

characteristic of the mitochondria themselves. This process is also independent of the presence or absence of magnesium in the medium, although this ion appears necessary to sustain valinomycin stimulated ATP-ase as reflected in the acid production during the final phase. This H^+ migration is most clearly seen in a system from which ATP has been omitted completely, and in Fig. 9 we view the process directly, uncomplicated by ATP-ase. Initially H^+ just pours out of the mitochondria, followed by a phase in which H^+ apparently returns to the mitochondria, followed by the third phase during which no further changes take place. If at this point we add DNP, we return immediately to the original pH. As a matter of fact, prior inclusion of DNP would have completely prevented any valinomycin induced pH effect. The amount of H^+ which valinomycin stimulated mitochondria can eject is of the order of 50 μmol/gm protein for a rat liver derived preparation. About one quarter of this value is obtained for pigeon heart preparations. The striking feature of the process is that valinomycin obviously plays a catalytic role, since as much as 1000 moles of H^+ can be expelled per mole valinomycin added. In many respects this phenomenon closely resembles the H^+ production associated with the binding of divalent metal ions by mitochondria (16-18). Both processes are substrate dependent. If antimycin is added to rat liver mitochondria to inhibit endogenous respiration, the H^+ expulsion process is also inhibited, while with a preparation of low endogenous substrate, such as pigeon heart mitochondria, no H^+ expulsion takes place unless substrate is provided. This is a true migration of H^+ across the mitochondrial membrane rather than a cyclic net production and consumption of H^+, since, if triton is added to the system at any point of the process so that the mitochondria dissolve and form a one phase system, the pH is immediately restored to the initial value. We do not yet know what is the source of the H^+ expelled from the mitochondria, but whatever it is there must be a large amount of it present. This does call attention to a complicating factor in interpreting rigorously the ratio of H^+ appearance to divalent metal binding, since other processes in the mitochondria, which may be triggered catalytically by metal ions, could also contribute to the net extramitochondrial H^+ increase observed.

While our general approach to the problem of energy transfer differs from the "fragmatic" approach alluded to in our introduction, it is of interest to speculate how the site specific reagents we have catalogued would affect the properties of various submitochondrial preparations and soluble factors. In our laboratory Dr. Hommes has already demonstrated that the differences in inhibitory action between DBI and octylguanidine vis-a-vis cytochrome b are discernible in one type of submitochondrial system (19). Our studies also render possible certain unique information about intact mitochondria, such as the inference we draw from the data presented here that parallel, non identical energy transfer pathways actually do exist in intact mitochondria prior to disruption.

REFERENCES

1. Lardy, H. A., and Elvejhem, C. A., Ann. Rev. Biochem., 14, 1 (1945).
2. Lardy, H. A., Johnson, D., and McMurray, W. C., Arch. Biochem. Biophys., 78, 587 (1958).
3. Hollunger, G., Acta Pharmacol. Toxicol., II, Suppl. 1 (1955).
4. Pressman, B. C., J. Biol. Chem., 238, 401 (1963).
5. Lardy, H. A., First IUB/IUBS Symposium, Biological Structure and Function, Stockholm, Academic Press Inc., New York 1962, p. 261.
6. Pressman, B. C., Fed. Proc., 21, 55 (1962).
7. Chance, B., and Hollunger, G., J. Biol. Chem., 236, 1534 (1961).
8. Chance, B., and Hollunger, G., Fed. Proc., 20, 50 (1961).
9. Heytler, P. G., and Prichard, W. W., Biochem. Biophys. Res. Comm., 7, 272 (1962).
10. Hemker, H. C., Biochim. Biophys. Acta, 63, 46 (1962).
11. Johnson, R. B., Feldott, G., and Lardy, H. A., Arch. Biochem. & Biophys., 28, 317 (1950).
12. McMurray and Begg, R. W., Arch. Biochem. & Biophys., 84, 546 (1959).
13. Schemjakin, M. M., Angew. Chem., 72, 342 (1960).
14. Bruni, A. L., Contessa, A. R., and Luciani, S., Biochim. et Biophys. Acta, 60, 301 (1962).
15. Packer, L., and Jacobs, E. E., Biochim. Biophys. Acta, 57, 371 (1962).
16. Chappell, this colloquium.
17. Brierley, this colloquium.
18. Vasington, F. D., and Murphy, J. V., J. Biol. Chem., 237, 2670 (1962).
19. Hommes, F. A., Biochim. Biophys. Acta, 8, 248 (1962).

DISCUSSION

Racker: Dr. Pressman, I wonder whether some of the differences in the action of these uncouplers may be linked to the structural relationships within the sites in intact mitochondria? In submitochondrial particles I could not observe the different effects on Site 1 and Site 2 which you have described for intact mitochondria.

Pressman: The development of the guanidine inhibitions seems to depend on the building up of a critical level of some energized intermediate. This is very obvious in the case of DBI, less apparent in the case of octylguanidine. Now it is possible that the actual levels which build up in a given particle preparation do not reach this critical level.

Racker: On your recommendation, I have tried DBI and dicoumarol at various concentrations. In all cases we uncoupled Site 1 when we used enough to uncouple Site 2.

Pressman: We really don't know how to predict what to expect in a submitochondrial system. The impression we would like to convey is that the analogous agents we have discussed do not al all act in identical manners. In the absence of guanidine derivatives all the uncouplers appear to have very similar effects, even on intact mitochondria, which we attribute to a rather free communication between the separate branches of the energy transfer pathways. It is only by interrupting this free communication by means of site-specific energy blocking agents that we are able to demonstrate that the uncoupling agents are also site-specific. I don't know anything firsthand about the properties of your particles, so I don't know the extent to which these considerations apply to them.

Chance: Do you need to bring in the energy levels? There certainly can be specificity changes. You have already shown that the oligomycin-sensitizing factor can be taken out; why can't some of these earlier factors come out as well?

Chappell: I assume that you consider the site of action of uncoupling agents to be between the guanidine-site and the site of interaction with ADP because the alkylguanidines do not affect DNP-stimulated ATP-ase. But surely, in the normally accepted scheme for ATP hydrolysis through the oxidative phosphorylation system:-

$$\text{ATP} + \text{X} \rightleftharpoons \text{ADP} + \text{X} \sim \text{P} \quad \ldots\ldots\ldots\ldots (1)$$
$$\text{X} \sim \text{P} + \text{I}_{1,2,3} \rightleftharpoons \text{X} \sim \text{I}_{1,2,3} + \text{P}_i \quad \ldots\ldots\ldots\ldots (2)$$
$$\text{X} \sim \text{I}_{1,2,3} \xrightarrow{\text{(DNP)}} \text{X} + \text{I}_{1,2,3} \quad \ldots\ldots\ldots\ldots (3)$$

if the rate-limiting reaction were (1), and there were three $X \sim I$ compounds, only the one involved at the level of the DPN-phosphorylation site ($X \sim I_1$) combining with the alkylguanidine, then one would expect no effect on DNP-stimulated ATP-ase, as has been suggested previously (1). Also, the apparent competition between alkylguanidine and DNP in respiration suggests that this is the correct interpretation.

<u>Pressman:</u> We have reservations in accepting the criteria cited in your paper as preemptive in establishing the sequence or identity of loci of action of DNP and the guanidines. Our basic reasoning is this: consider a system which is strongly inhibited by alkylguanidine, as evidenced by low levels of phosphorylation and respiration in the presence of ADP. We know that respiration would resume after addition of DNP only after a significant time delay while maximal ATP-ase activity following the addition of DNP is instantaneous. We have confirmed your observation that the maximal phosphorylation rates attainable are approximately equal to the maximal rate of DPN-stimulated ATP-ase. It does not seem likely that alkylguanidines block the energy transfer pathway at the locus of DNP action by occupying this site, while the alkylguanidines fail to produce even a transient blocking of DNP from this site in the ATP-ase test system.

Note added in proof:

We will concede that the point Dr. Chappell has raised renders our argument less conclusive and open to reappraisal if further experimentation should so warrant. But for the following reasons we feel that the evidence cited for competition between alkylguandines and DNP is at least equally inconclusive.

1. Guanidine inhibition is antagonized not only by DNP and other uncoupling agents, but also, as we have mentioned, by partial substrate inhibition (malonate) and according to Chance and Hollunger (2) even by ADP. The latter investigators even find a similar antagonism between Amytal and the uncoupler DIB (3). It would be impossible for all these agents to function at the same locus and the assembled data suggest rather that the antagonism is indirect and mediated through lowering the level of one or more intermediates of the energy transfer sequences upon which the guanidine inhibition is

dependent.

2. We have further analyzed the kinetics of DNP reversal of octylguanidine respiratory inhibition and find that the protracted lag phase (cf. Fig. 2 of my paper, especially the 60 µM DNP tracing) is inconsistant with a simple, direct octylguanidine-DNP competition.

3. An alkylguanidine-DNP direct competition fails to explain the incomplete recoveries of respiration we obtain with DNP or other uncouplers, in the presence of high levels of alkylguanidines.

Chappell: Concerning your point that DNP does not instantly reverse the inhibitory effect of alkylguanidine on oxidative systems which are NAD-linked, surely this indicates only that the DNP is able to displace the guanidine only slowly? The combination of the alkylguanidines with whatever it is they do combine with ($?X \sim I_1$) is after all a very slow process.

To take your second point, I am glad that you have been able to confirm the equality of the rates of DNP-stimulated ATP-ase and the maximal rate of phosphorylation; this obtains not only with liver mitochondria but also with heart sarcosome However, I don't see that this supports your contention. It may indicate only that reaction (1) is rate-limiting in both directions; it could be for example that the access of adenine nucleotide to X is the rate-controlling step.

In reply to your third point, I agree that a precise kinetic analysis of the DNP-alkylguanidine competition is made difficult by the slowness of the reactions. However, from your scheme I am unable to see how DNP could ever reverse the inhibition caused by the guanidines.

Green: I am amazed at the magnitude of the acid migration, presumably from the inside to the outside, which you could not correlate with the accumulation of any of the divalent salts because they were not there--is that correct?

Pressman: That is right.

Green: The only conditions under which we can find hydrogen ion production are those in which ions such as magnesium and calcium are accumulated. If there is no movement of Mg^{++} or Ca^{++} from the outside to the inside of the mitochondrion, then there must be discovered a device by which H^+ can be translocated independent of such movement. The picture that emerges from our studies is that it is the deposition of calcium phosphate or magnesium phosphate inside the mitochondrion, which leads to the production of H^+. There must be some equivalent process operative under your conditions.

Pressman: We have reservations as to whether or not precipitation of a salt is a necessary requisite for the ejection of H^+ from mitochondria. I think Dr. Chappell touched on some of this today.

Green: But at any rate, I cannot see the relationship of your system to other systems in which H^+ is produced. It would appear to be an extremely novel mechanism.

Pressman: The ejection of H^+ from mitochondria triggered by valinomycin has this in common with the metal ion triggered process: it is inhibited by other uncoupling agents such as dinitrophenol, it requires substrate-derived energy, but it is not inhibited at all by oligomycin. The two processes have a great deal in common. It seems notable that the H^+ migration stimulated by valinomycin takes place in the opposite direction to that invoked by other uncoupling agents.

Chance: Could you explain in more detail what you mean by "opposite direction?"

Pressman: When you take the buffered system and add the usual uncoupling agents such as DNP, you actually get a rise in the extramitochondrial pH. This H^+ migration with DNP is presumably the process Peter Mitchell originally observed (4). The H^+ shift with valinomycin is in the direction opposite to that produced by all other uncoupling agents tested except gramacidin. Valinomycin shares all properties that we have observed, at least qualitatively, with gramacidin.

Green: That is not quite correct. You get exactly the same direction of H^+ shift when mitochondria accumulate either Mg^{++} or Ca^{++}. The only difference is that there is no translocation of divalent metal ions in your system.

Pressman: A significant difference between the H^+ translocation triggered by metal ions, and that triggered by valinomycin is that in the latter case no simple stoichiometry obtains between the agent added to the mitochondria and the extent of the H^+ ejection produced.

REFERENCES

1. Chappell, J. B., J. Biol. Chem., 238, 410 (1963).
2. Chance, B., and Hollunger, G., J. Biol. Chem., 278, 432 (1963).
3. Chance, B., and Hollunger, G., J. Biol. Chem., 278, 418 (1963).
4. Mitchell, P., Biochem. J., 81, 24 p. (1961).

THE UTILIZATION OF REDUCING POWER

RELATIONSHIPS BETWEEN ENERGY-GENERATION AND NET ELECTRON TRANSFER IN BACTERIAL PHOTOSYNTHESIS

Subir K. Bose* and Howard Gest

The Henry Shaw School of Botany and
The Adolphus Busch III Laboratory of Molecular Biology
Washington University, St. Louis, Missouri

One of the central problems of bacterial photosynthesis is the mechanism(s) by which _net_ reducing power is generated. It is our view that this aspect of photosynthetic metabolism may differ fundamentally in the bacteria as compared with green plants. The present paper summarizes evidence which is consistent with the hypothesis (1-3) that net electron flow in bacterial photosynthesis occurs, in part, through "reverse" electron transfer mechanisms similar to those observed in mitochondria.

An outstanding difference between the autotrophic metabolism of green plants and the anaerobic photosynthetic bacteria is found in the fact that the bacteria require, in addition to CO_2 and water, an oxidizable inorganic compound - frequently referred to as the "accessory hydrogen (or electron) donor" (4). Many photosynthetic bacteria also have the capacity to grow as anaerobic "autoheterotrophs" (5), _i.e._, using light as the energy source and a single organic compound as the ultimate source of carbon and reducing power. Various inorganic accessory donors or organic compounds utilized by photosynthetic bacteria would be expected, in principle, to provide a source of $NADH_2$ (or $NADPH_2$) through dark dehydrogenase activity. With donors of relatively high redox potential, however, the formation of $NADH_2$ presumably requires intervention of the photochemical apparatus. According to one popular hypothesis (6), designated as "noncyclic electron flow", NAD is reduced by electrons derived from photoexcited bacteriochlorophyll (bchl.), while the high potential donor supplies

---

This investigation was supported by research grants from the National Science Foundation (G-9877) and the U.S. Public Health Service (E-2640).

*Trainee supported by U.S. Public Health Service Training Grant 2G-714.

electrons for reduction of the oxidized pigment to its original state. Some investigators (7, 8) assume that net formation of $NADH_2$ occurs through a noncyclic mechanism of this kind with all accessory donors, regardless of redox potential.

The discovery of light-dependent anaerobic phosphorylation (9) in photosynthetic bacteria and energy-linked electron transfer in mitochondria provide the basis for a different, and perhaps more plausible, explanation of light-stimulated net electron transfer in bacterial photosynthesis. Subcellular pigmented particles from the purple bacterium Rhodospirillum rubrum catalyze light-dependent phosphorylation of ADP in the absence of significant quantities of added electron donors or acceptors. In addition, they can effect an energy-dependent reduction of NAD by succinate, in the sense that a net reaction occurs only when the system is illuminated (10). Although the light-dependency of this oxidation-reduction process can be interpreted on the basis of the noncyclic hypothesis noted above, it is also possible that light promotes the reaction by providing energy-rich intermediates (i.e., precursors of ATP, generated during light-induced "cyclic" electron transfer) which drive hydrogen transfer from succinate to NAD. The latter interpretation receives some support from the observation (10, 11) that the succinate-dependent photoreduction of NAD is markedly inhibited when optimal quantities of $ADP + P_i + Mg^{++}$ are added to the system. This inhibition may well result from a competition between the phosphorylation and oxidation-reduction systems for an energy-rich compound required for both processes.

In view of the various metabolic similarities between mitochondria and the pigmented particles from R. rubrum, it is of some interest to comment briefly on the in vivo organization of the bacterial photochemical apparatus. Several types of evidence (12, 13) indicate that in Rhodospirillum the photochemical system and its associated enzymes, and electron carriers, are integrated with the cytoplasmic membrane, or intracytoplasmic extensions of the membrane. Disruption of such membranes by any one of a number of drastic cell-breakage procedures now in common use (e.g., sonic oscillation, alumina grinding, French pressure cell) appears to result in formation of the pigmented photoactive particles ("chromatophores"). Although our knowledge is relatively limited as yet, it seems that the energy-generating system of R. rubrum is embedded in a complex membranous structure which is comparable in a general way to that of the mitochondrion.

Light-dependent reduction of fumarate in a model system

TABLE I

Effect of light on reduction of fumarate with H_2 by R. rubrum particles: relationship with E_o' of the redox mediator.

Mediator	E_o' (mv.)	Role of Light
1. Benzyl viologen	-359	none
2. Janus Green B	-256	none
3. Flavin mononucleotide*	-219	none
4. Flavin adenine dinucleotide*	-219	none
5. Nile Blue A	-119	none
6. Menadione (Vitamin K_3)	ca. -44	none
7. Methylene Blue	+ 11	stimulatory
8. Brilliant Cresyl Blue	+ 47	stimulatory
9. Thionine	+ 63	stimulatory
10. N-methyl phenazonium methosulfate	+ 80	stimulatory
11. Coenzyme Q_{10} (R. rubrum)	+ 89	stimulatory
12. Toluylene blue	+115	stimulatory
13. 1-naphthol-2-sulfonate-DCPIP	+119	obligatory
14. 2,6-dichlorophenol-indophenol	+217	obligatory
15. Tetramethyl-p-phenylenediamine	+260	obligatory
16. Cytochrome c, mammalian	+262	obligatory
17. p-Benzoquinone	+280	obligatory
18. Rhodospirillum cytochrome c_2	+330	obligatory

*Mediator function not observed in the absence of a detergent such as deoxycholate (see ref. 3).

Particle preparations obtained from hydrogen-producing cells of R. rubrum contain an active hydrogenase which is capable of reducing a number of electron acceptors with H_2 in darkness. Although fumarate is not reduced by unsupplemented particles, H_2 is oxidized with this acceptor if a catalytic quantity of a suitable redox carrier is also added (3). With redox mediators of low potential (relative to that of the succinate-fumarate couple; E_o' = +31 mv.), hydrogenase activity with fumarate is observed in the dark and illumination has no significant effect (Table I). When the mediator has a moderately high potential ($\sim$0 to +115 mv.), the reaction rate is stimulated by light. With mediators of even higher potential, the reduction of fumarate with H_2 is dependent on illumination. Fumarate reduction in such systems is not appreciably inhibited by excess malonate (e.g., even with malonate/fumarate = 10) and it may be noted that insensitivity to malonate appears to be a characteristic feature of bacterial fumaric reductase activity (14).

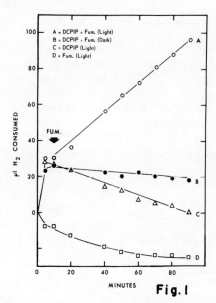

Fig. 1

<u>Light-dependent reduction of fumarate with H_2 mediated by DCPIP</u>. R. rubrum particles were obtained from cells grown photosynthetically in a succinate + glutamate medium as described in ref. 15. The reaction mixtures contained 100 μmoles of potassium phosphate buffer pH 7.4 and particles equivalent to 0.59 mg bacteriochlorophyll. Where indicated, one side-arm of the Warburg vessel contained 1.5 μmoles of DCPIP and the second side-arm, 20 μmoles of sodium fumarate. The final fluid volume was 2.0 ml and 0.2 ml of 20% KOH was present in the center well. Gas phase, H_2; temperature, 30° C; light intensity, 300 foot-candles. In the complete system, the dye was reduced by the hydrogenase and fumarate added subsequently.

The requirements for demonstration of the light-dependent fumarate reduction activity are shown by the data of Fig. 1. In this experiment, the high redox potential mediator was 2,6-dichlorophenolindophenol (DCPIP; E_o = +217 mv.), which is rapidly reduced by the R. rubrum hydrogenase. It is evident that following reduction of the dye, further H_2 consumption is not observed unless both fumarate and light are provided. The mediator remains in the reduced (leuco) form during the course of the reaction (curve A) owing to the presence of excess hydrogenase. Similar results are obtained using N,N,N', N'-tetramethyl-p-phenylenediamine (TMPD; E_o' = +260 mv.) as the mediator, but in this case H_2 is not consumed prior to fumarate addition.

It is noteworthy that particles stored in concentrated suspension in 0.005 M potassium phosphate buffer pH 7.6 at $0°$ C. under a gas phase of hydrogen retain activity for at least several weeks. The light-dependent activity is not inhibited by chelating agents such as ethylenediaminetetraacetic acid and addition of Mg^{++} or other divalent cations is not required. With aged particles, however, 5 mM Mg^{++} or Mn induces appreciable activity in darkness (with either DCPIP or TMPD), and the reaction rate is stimulated about 10-fold by illumination.

The fact that illumination stimulates or is required only when the added catalyst has a high redox potential relative to that of the succinate-fumarate couple suggests that the action of light is concerned with generating energy-rich intermediates which promote hydrogen transfer against an unfavorable thermodynamic gradient, i.e., from reduced dye to fumarate. This interpretation is supported by the results of a number of experiments in which it was observed that the overall oxidation-reduction reaction was markedly inhibited by addition of the reactants required for photophosphorylation (i.e., ADP + P_i + Mg^{++}). For reasons still unknown, however, this effect-which is reminiscent of the inhibition of succinate-dependent photoreduction of NAD phosphorylation reactants-is not shown by all particle preparations. Attempts to demonstrate promotion of fumarate reduction with H_2 in the presence DCPIP or TMPD by added ATP, in the dark, have given negative results thus far.

Effects of uncoupling agents on light-dependent fumarate reduction

A number of agents which uncouple oxidative phosphorylation in mitochondria have been tested for their effects on fumarate reductase activity, with TMPD serving as the mediator, in the presence of the photophosphorylation reactants. The results are summarized in Tables II and III.

TABLE II

Inhibition of light-dependent fumarate reduction with H_2 and phosphorylation by certain uncoupling agents

Inhibitor	Inhibitor conc. (M)	Particle conc. ($\frac{mg\ bchl.}{3ml}$)	Specific Activity (μmoles/hr./mg bchl.) H_2 oxidation	Phosphorylation
Dicumarol	0	0.21	46.0	457.0
	5×10^{-4}	"	11.9	14.3
	7×10^{-4}	"	7.6	11.5
	1×10^{-3}	"	0	0
PCP*	0	0.58	11.0	175.5
	1×10^{-4}	"	7.7	175.5
	2.5×10^{-4}	"	4.4	175.5
	5×10^{-4}	"	1.3	79.0
	7.5×10^{-4}	"	0	3.5
m-CCP**	0	0.45	16.7	290.0
	5×10^{-5}	"	5.4	0
	1×10^{-4}	"	0	0
DIB***	0	0.18	47.8	572.8
	1×10^{-4}	"	34.4	448.0
	5×10^{-4}	"	8.4	25.6

*   pentachlorophenol
** m-chlorocarbonylcyanide phenylhydrazone
*** n-butyl-3,5-diiodo-4-hydroxybenzoate

The reaction mixtures (in Warburg vessels) contained, in a final volume of 3.0 ml: twice washed R. rubrum particles, as indicated; Tris-HCl pH 7.9, 200 μmoles; TMPD, 2 μmoles; sodium fumarate, 40 μmoles; ADP, 2 μmoles; K_2HPO_4, 35 to 50 μmoles; $MgCl_2$, 30 μmoles; hexokinase, 1.5 mg; mannose, 50 to 60 μmoles. Gas phase, H_2 (KOH in center well); temperature, 30° C; light (red) intensity, 2000 foot-candles. The particles were preincubated with inhibitors for approximately 10 minutes and the extent of phosphorylation determined in the usual way after deproteinization of the suspensions used for the manometric assays.

It is evident that Dicumarol, PCP, m-CCP, and DIB inhibit both photophosphorylation and the light-dependent reduction of fumarate with H_2. Other experiments showed that the degree of inhibition by such agents is greatly influenced by the concentration of R. rubrum particles used. On the other hand, dark reduction of fumarate with H_2 mediated by a low potential dye (Nile Blue A; E_0' = -119 mv.) is not inhibited by the compounds listed in Table II, even at concentrations which completely abolish the light-dependent activity. These results suggest that with high redox potential mediators the reduction of fumarate requires energy-rich intermediates produced by the photochemical system; dissipation of the postulated intermediates presumably would lead to inhibition of the overall reaction (16). Interference with formation of such energy-rich intermediates by inhibitors of electron transport also would be expected to have the same effect - phenylmercuric acetate (PMA; an inhibitor of light-induced electron transfer (17)) and trifluorothienylbutanedione (TTB; which inhibits reduction of coenzyme Q in mitochondrial preparations (18)) do, in fact, inhibit both photophosphorylation and the light-dependent reduction of fumarate.

TABLE III

Effect of oligomycin on light-dependent fumarate reduction with H_2 and photophosphorylation

Oligomycin conc. (M)	Specific Activity (μmoles/hr./mg bchl.)	
	H_2 oxidation	Phosphorylation
0	15.0	225.0
9×10^{-7}	17.7	71.6
3×10^{-6}	18.8	31.0
6×10^{-6}	18.8	0
9×10^{-6}	17.2	0

Conditions as in Table II, except that the R. rubrum particles in each vessel contained 0.56 mg bchl.

As in energy-linked mitochondrial processes (reverse electron transfer, ion accumulation) light-dependent fumarate reductase activity in the TMPD model system is not inhibited by oligomycin. In fact, a slight stimulation of the reaction is observed (Table III), even though net phosphorylation is markedly suppressed (cf. 19).

Similar results, i.e., inhibition of phosphorylation but not of fumarate reduction, have been observed with both atebrin and Gramicidin D. In this connection it is of interest that atebrin inhibits the ATPase and ATP-P_i^{32} exchange activities of R. rubrum particles (20). On the basis of analogous studies (21, 22) with mitochondria, it seems reasonable to conclude that inhibitors of this kind do not prevent the generation of energy-rich intermediates by the photochemical apparatus but, rather, inhibit a phosphate transfer reaction involved in the terminal stages of phosphorylation of ADP.

Photoproduction of molecular hydrogen

Photoproduction of H_2 by purple bacteria is a particularly interesting aspect of light-dependent net electron flow. The formation of H_2 by R. rubrum is dependent on light and, except for cells which contain readily expendable reserves, on the addition of oxidizable organic substrates, e.g., citric acid cycle intermediates (4, 5, 23, 24). Under appropriate conditions, truly resting cells can completely dissimilate such compounds to CO_2 and H_2. This remarkable metabolic conversion is apparently due to the extensive operation of an anaerobic citric acid cycle, coupled with an additional light-dependent process which in some way effects the oxidation of reduced pyridine nucleotide by liberation of H_2 (23, 24).

The evidence available at present suggests that photoproduction of H_2 characteristically occurs when the supplies of energy and reducing power exceed the demands of the biosynthetic machinery of the cell. Accordingly, the H_2-producing anaerobic citric acid cycle is interpreted to represent a regulatory mechanism which maintains ATP and $NADH_2$ at levels consistent with the overall rate of biosynthetic activity (5, 23, 24). This view of the metabolic significance of H_2 production by purple bacteria is reinforced by the results of experiments with certain compounds which inhibit the photophosphorylation system. Thus, photoproduction of H_2 is inhibited by antimycin A, 2,4-dinitrophenol (DNP), and various redox dyes (24). Oligomycin, on the other hand, has no significant effect. Typical results with this antibiotic and effective inhibitors are shown in Fig. 2

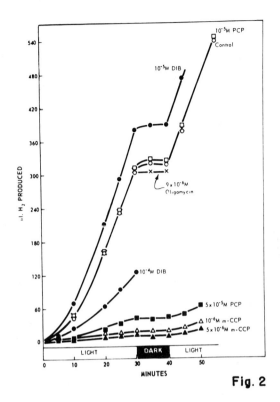

Fig. 2

Effects of uncoupling agents on photoproduction of H_2 from L-malate by intact cells of R. rubrum. The Warburg vessels contained, in a final volume of 2.0 ml: cells grown photosynthetically in a 0.6% DL-malate + 0.1% L-glutamate medium (24), equivalent to 1.2 mg cell nitrogen; potassium phosphate buffer pH 6.8, 100 μmoles; L-malate, 20 μmoles (added at zero time); other additions, as indicated. Gas phase, helium (KOH in center well); temperature, 30° C.; light (red) intensity, 2000 foot-candles.

Concluding remarks

Our current working hypothesis (3, 5) for net electron flow in bacterial photosynthesis is summarized in Fig. 3.

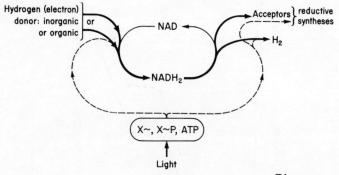

Fig. 3

<u>Scheme for net electron flow in bacterial photosynthesis.</u> The pyridine nucleotide may be either NAD or NADP; for convenience only the former is shown.

According to this scheme, electrons (or hydrogen) required for net reduction of pyridine nucleotide are derived from an accessory inorganic electron donor or organic substrates. As indicated, formation of $NADH_2$ may in some instances be a dark process and, in others, may be driven by energy-rich intermediates supplied by the photochemical apparatus. Reduced pyridine nucleotide would be used for reductive syntheses or, under conditions of limited biosynthetic activity, would be reoxidized through liberation of H_2 by an energy-dependent mechanism.

It may be suggested that at least two energy-rich intermediates are generated by the photochemical system of <u>R. rubrum</u> and similar bacteria, as depicted below.

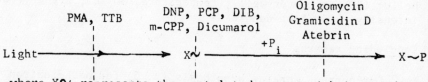

-where $X\sim$ represents the postulated energy-rich intermediate which can: (a) drive endergonic processes such as reverse electron transfer, production of H_2, and ion transport (25), and (b) be converted to $X\sim P$, which is utilized for phosphorylation of ADP.

Analogous schemes, in which intermediates of oxidative phosphorylation promote reverse electron transfer could explain, in principle, how certain aerobic chemosynthetic autotrophs generate reduced pyridine nucleotides with electron donors of relatively high redox potential. It is to be hoped that the present results will stimulate further study of electron transfer mechanisms in photosynthetic bacteria and non-photosynthetic autotrophs with the approaches and techniques now being used so successfully in the exploration of energy-linked processes in mitochondria.

REFERENCES

1. Chance, B. and Olson, J.M., Arch. Biochem. Biophys., 88, 54 (1960).
2. Chance, B. and Nishimura, M., Proc. Natl. Acad. Sci. U.S., 46, 19 (1960).
3. Bose, S.K. and Gest, H., Nature, Lond., 195, 1168 (1962).
4. Gest, H. and Kamen, M.D., in Ruhland, W. (Editor) Encyclopedia of Plant Physiology, Vol. V/2, Springer-Verlag, Berlin-Gottingen-Heidelberg, 1960, p. 568.
5. Gest, H., C.F. Kettering Foundation Symposium on Bacterial Photosynthesis, 1963, in press.
6. Arnon, D.I., Losada, M., Mozaki, M., and Tagawa, K., Nature, Lond., 190, 601 (1961).
7. Gaffron, H., in Kasha, M. and Pullman, B. (Editors), Horizons in Biochemistry, Academic Press, N.Y., 1962, p.59.
8. van Niel, C.B., Ann. Rev. Plant Physiol., 13, 1 (1962).
9. Frenkel, A.W., J. Amer. Chem. Soc., 76, 5568 (1954).
10. Frenkel, A.W., Brookhaven Symp. in Biol., 11, 278 (1959).
11. Horio, T., Yamashita, J. and Nishikawa, K., Biochim. Biophys. Acta, 66, 37 (1963).
12. Tuttle, A.L. and Gest, H., Proc. Natl. Acad. Sci. U.S., 45, 1261 (1959).
13. Giesbrecht, P. and Drews, G., Arch. Mikrobiol., 43, 152 (1962).
14. Peck, H.D., Jr., Smith, O.H. and Gest, H., Biochim. Biophys. Acta, 25, 142 (1957).
15. Bose, S.K. and Gest, H., Proc. Natl. Acad. Sci. U.S., 49, 337 (1963).
16. Chance, B. and Hollunger, G., J. Biol. Chem., 238, 445 (1963).

17. Nishimura, M. and Chance, B., Biochim. Biophys. Acta, 66, 1 (1963).

18. Ziegler, D.M., Biological Structure and Function, Proc. First IUB/IUBS Symp., Academic Press, London-New York, Vol II, 1961, p. 253.

19. Baltscheffsky, H. and Baltscheffsky, M., Acta Chem. Scand. 14, 257 (1960).

20. Baltscheffsky, H. and Baltscheffsky, M., Acta Chem. Scand. 12, 1335 (1958).

21. Ernster, L., Biological Structure and Function, Proc. First IUB/IUBS Symp., Academic Press, London-New York, Vol. II, 1961, p. 139.

22. Slater, E.C., Tager, J.M., and Snoswell, A.M., Biochim. Biophys. Acta, 56, 177 (1962).

23. Ormerod, J.G. and Gest, H., Bact. Rev., 26, 51 (1962).

24. Gest, H., Ormerod, J.G., and Ormerod, K.S., Arch. Biochem. Biophys., 97, 21 (1962).

25. Bose, S.K., Gest, H., and Ormerod, J.G., J. Biol. Chem., 236, PC 13 (1961).

THE ACCUMULATION OF DIVALENT IONS BY ISOLATED MITOCHONDRIA

J.B. Chappell, Mildred Cohn and G.D. Greville

The Department of Biochemistry, University of Cambridge,
The Johnson Foundation, University of Pennsylvania, and
The Biochemistry Department, Institute of Animal Physiology,
Babraham, Cambridge

It is apparent from work originating in Madison (Brierley et al, 1962, 1963), Baltimore (Lehninger et al. 1963 a,b; Vasington and Murphy, 1962) and Cambridge (Chappell et al. 1962; Chappell and Greville, 1963) that isolated mitochondria accumulate divalent metal ions, including Ca^{++}, Mg^{++}, Mn^{++} and Sr^{++}, with the simultaneous uptake of P_i, in a respiration-dependent process.

THE ACCUMULATION OF Mn^{++} IN THE PRESENCE AND ABSENCE OF P_i

(The experiments reported in this section were carried out in collaboration with Dr. G.D. Greville (A.R.C. Babraham) and with the assistance of Miss M.J. Burroughs and Mr. K.E. Bicknell.)

In Table I are shown the results of a series of experiments in which Mn^{++} uptake, oxygen consumption and H^+ production were followed. The mitochondria were separated from the suspension media by filtration through an Oxoid membrane (Oxoid Division of Oxo Ltd., London, E.C.4; a "millipore" type membrane). H^+ production was followed with a recording pH-meter and O_2 consumption with a Clark oxygen electrode. P_i and Mn^{++} uptake were determined by assay of the filtrates by the method of Fiske and SubbaRow (1925) and from the ^{54}Mn content by determination of radioactivity, respectively.

This paper was written and delivered by J.B. Chappell.

TABLE I

Mn^{++} ACCUMULATION BY RLM IN THE PRESENCE AND ABSENCE OF PHOSPHATE

($^{54}Mn^{++}$-Rapid Filtration Technique)

pH 7.14
1.8 μmole-Mn^{++}

With 1 mM Pi

Mn^{++} uptake (μA)	Pi uptake (μmole)			P_i/Mn^{++}
1.60 (11)	1.09 (11)			0.68

Mn^{++} uptake (μA)	Extra O (μA)	Mn^{++}/O	H^+ produced (μeq.)	H^+/Mn^{++}
1.67 (26)	0.31 (26)	5.4	1.49	0.90

No added Pi

Mn^{++} uptake (μA)	H^+ produced (μeq.)			H^+/Mn^{++}
0.64 (35)	0.72 (35)			1.12

It will be seen from Table I that the P_i/Mn^{++} ratio was 0.68 (for $Mn_3(PO_4)_2$, the ratio is 0.67), the H^+/Mn^{++} ratio was 0.90, which is close to the value expected for the reactions:

$$3\ Mn^{++} + 2\ H_2PO_4^- \longrightarrow Mn_3(PO_4)_2 + 4H^+ \quad (1)$$
$$3\ Mn^{++} + 2\ HPO_4^{--} \longrightarrow Mn_3(PO_4)_2 + 2H^+ \quad (2)$$

at pH 7.15, since this pH value is close to that of the second dissociation constant of P_i.

Calculation of the Mn/O ratio was complicated by the uncertainties illustrated in Fig. 1. However, by determining the variation of extra O_2 uptake with varying amounts of Mn^{++} added, it was possible to derive a value from the slope of a plot of extra O_2 uptake against Mn^{++} added; this procedure gave a value of approximately 6 Mn^{++}/O. This value, when corrected for the uptake of Mn^{++} which occurred when respiration was inhibited (see below), was 5.4 in the series shown in Table I.

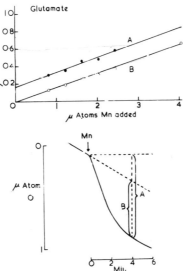

Fig. 1.
The lower curve (solid line) is a typical O_2 electrode trace obtained when Mn^{++} was added to a mitochondrial suspension in the presence of P_i. Extra O_2 uptake was calculated on basis A or B with the results shown in the upper part of the figure. The slope of curve A was 6.0 and B 6.2 µ atoms Mn^{++}/µ atom O. Glutamate was the substrate.

The accumulation of Mn^{++} and P_i and the associated production of H^+ were prevented by, (1) the respiratory inhibitors (HCN and antimycin inhibited with all substrates tested, amytal and rotenone inhibited with those substrates with NAD-linked dehydrogenases, e.g. glutamate-malate, β-hydroxybutyrate, but not with succinate); (2) DNP and 2,4-dibromophenol. In contrast oligomycin was completely without effect. In the presence of added P_i, once accumulation of Mn^{++} and P_i had occurred, the addition of uncoupling agents or respiratory inhibitors did not cause release of the accumulated Mn^{++}. Release of Mn^{++} and reversal of the pH effect, could be achieved only by adding lytic quantities of the detergent Triton X-100.

Once Mn^{++} had been accumulated by mitochondria in the presence of added P_i it was not available to added chelating agents, e.g. ethylene-glycol-bis-(aminoethyl)-tetracetate (EGTA). Free Mn^{++} reacts with EGTA thus:

$$\text{EGTA} + Mn^{++} \longrightarrow \text{EGTA} \cdot Mn^{++} + 2H^+ \quad \text{---------- (3)}$$

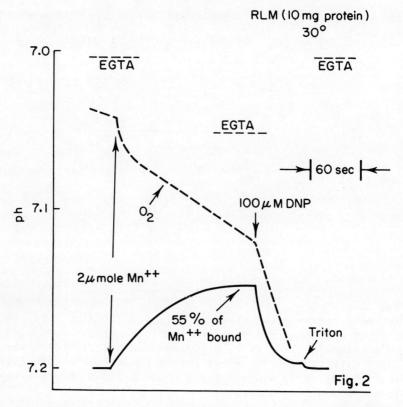

Fig. 2

Oxygen electrode trace (hatched curve) and pH-recording (solid curve) obtained when Mn^{++} was added to rat liver mitochondria in the absence of P_i. Total volume 4.0 ml., containing: 80 mM KCl, 20 mM Tris-Cl, pH 7.2, 5 mM glutamate, 5 mM malate and 10 mg mitochondrial protein. The pH changes obtained when 4.0 μmoles EGTA was added, in separate control experiments, are indicated.

Since the uptake of Mn^{++} by mitochondria produced only approximately one H^+ (see Table I) it was possible to determine the extent to which Mn^{++} was taken up by the mitochondria, at any instant, by adding an excess of EGTA and recording pH changes (see Fig. 2). This approach has been used to show that on addition of Ca^{++} or Sr^{++} to mitochondria, in the presence of P_i, there was an almost quantitative accumulation of these cations, as was the case with Mn^{++}. Again, these processes were respiration-dependent, DNP-inhibited and oligomycin-insensitive. The initial rates of respiration, induced by addition of divalent cations, DNP and ADP are compared in Table II.

TABLE II

INITIAL RATES OF OXYGEN CONSUMPTION INDUCED BY DIVALENT CATIONS ADP AND DNP.

Addition	Rate of oxygen consumption (μA O/mg protein/min)
0.2mM-ADP	0.105
0.1mM-DNP	0.160
0.25mM-CaCl$_2$	0.210-0.180
0.25mM-SrCl$_2$	0.110
0.5mM-MnSO$_4$	0.065

Rat liver mitochondria (10 mg protein) were added to a medium containing (final concentrations) 80 mM-KCl, 20mM-Tris-Cl, 5mM-P$_i$, 5mM-glutamate and 5mM-malate. Total volume 4.0 ml, temp. 30°. The Clark oxygen electrode was used to determine rates.

Ca^{++} induced rates of electron transport in excess of those produced by ADP (see Chance, 1959) and in this respect resembled DNP (Chappell, 1962). Sr^{++} produced rates of respiration very similar to those obtained with ADP, and Mn^{++}, rates about 60% of this. Preliminary results indicate that (in the case of the divalent metal ions) these differences are due to a variation of affinity rather than maximum velocity.

P_i could be replaced by arsenate in these experiments when Mn^{++} and Sr^{++} were used. However, with Ca^{++}, swelling occurred almost immediately in the presence of arsenate and this prevented uptake of Ca^{++} (see below).

Mn^{++} was also accumulated by rat liver mitochondria in the absence of P_i or arsenate (see Table I, Figs. 2 and 3). Again, this process was respiration-dependent (i.e. appropriate respiratory inhibitors prevented the effect), DNP inhibited, but oligomycin was without effect. Between 200-300 μA Mn^{++}/gm protein could be bound by the mitochondria in this way. 1 mg/ml bovine plasma albumin stablilsed the system. The accumulation differed from that in the presence of P_i in that DNP both inhibited and <u>reversed</u> the accumulation of Mn^{++} i.e. if DNP was added after Mn^{++} had been accumulated, the Mn left the mitochondria (Fig. 2 and 4). Respiratory inhibitors, e.g. antimycin, also caused release of Mn^{++} which had been accumulated by mitochondria in the absence of P_i, but this release occurred far more slowly than that induced by DNP. If P_i was added immediately after the antimycin, then

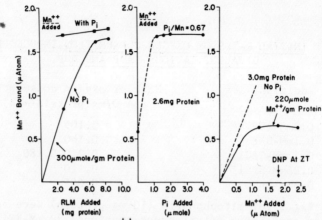

Fig. 3. The uptake of Mn^{++} in the presence and absence of P_i. Basic conditions: 80 mM KCl, 20 mM Tris-Cl, pH 7.2, 5 mM glutamate, 5 mM malate. (A) 0.45 mM $MnSO_4$, 1 mM P_i when present. (B) 0.45 mM $MnSO_4$, 2.6 mg mitochondrial protein. (C) 3.0 mg protein.

---

most of the accumulated Mn^{++} stayed in the mitochondria and P_i was taken up in a <u>respiration independent process</u>.

It appears, therefore, that the primary process is that of the respiration-dependent accumulation of the divalent metal cation and that the uptake of P_i (or arsenate) is a secondary process.

<u>INVESTIGATION OF THE Mn^{++}-EFFECT USING EPR AND PULSED-NMR</u>

(These investigations were carried out in collaboration with Dr. Mildred Cohn in the Johnson Foundation, University of Pennsylvania. Relaxation rates were kindly determined by Mr. J. Leigh.)

Cohn and her co-workers have pioneered the use of the technique of following the proton relaxation rate (PRR) of water for the study of the binding of divalent metal ions to proteins (see Cohn, 1963). Mn^{++}, because it possesses five unpaired-electrons, has a marked effect on the relaxation rate of water. It has been shown that it is the water which is co-ordinated to the Mn^{++} which is affected. However, if the Mn^{++} is bound to a macromolecule at a site which permits accessibility of the water to the Mn^{++}, then the relaxation rate is usually increased. This is called enhancement (ϵ_b) and the extent of this is given by:

$$\epsilon_b = \frac{\text{Relaxation rate observed with bound } Mn^{++}}{\text{Relaxation rate when } Mn^{++} \text{ is not bound}} \quad \text{------ (4)}$$

Typical values which have been observed are:
(1) ϵ_b = 5-30, for binding to some enzymes, bovine plasma albumin, desoxyribonucleic acid.
(2) ϵ_b = ca.0.5, for Mn^{++} bound by EDTA (King and Davidson, 1958); in this complex all the co-ordination positions are bound.

The experiments described below were carried out in an attempt to determine the course of uptake of Mn^{++} by liver mitochondria. The uptake of Mn^{++} was followed simultaneously by EPR (to determine the amounts of free and bound Mn^{++}) and by pulsed-NMR (to determine the proton relaxation rate). From these measurements the enhancement of the bound Mn^{++} was determined.

In Fig. 4 are shown the results of an experiment in which Mn^{++} uptake by mitochondria was followed by EPR. The following points should be noted:
(1) Initially, there was a rapid binding of Mn^{++} to the mitochondria. This was <u>not</u> inhibited by DNP or by antimycin or rotenone.
(2) In the <u>presence</u> of P_i there was a rapid binding of Mn^{++}. This was prevented by addition of DNP at zero-time, but not reversed by adding DNP subsequently.
(3) In the absence of P_i, Mn^{++} was accumulated by the mitochondria, less rapidly and less extensively. DNP prevented the accumulation and reversed it, once it had occurred.
(4) In all cases, addition of Triton caused release of Mn^{++}; the amount of free ion was very nearly the same as the zero-time, extrapolated concentration. These results confirm and extended those obtained by assay of Mn^{++} by the rapid filtration technique.

Calculation of the enhancement of relaxation rate of the bound Mn^{++} at various stages gave the following results:

(1) The initial, rapid, DNP-, antimycin- and oligomycin-insensitive binding of Mn^{++} gave an ϵ_b value of 5-7. This would be the case if the Mn^{++} were bound to protein or nucleic acid. It is suggested that this stage represents a non-specific "surface" binding of Mn^{++}, which is possibly not associated with the accumulation process per se, since DNP, and respiratory inhibitors, had no effect. Both Mg^{++} and Ca^{++} competed

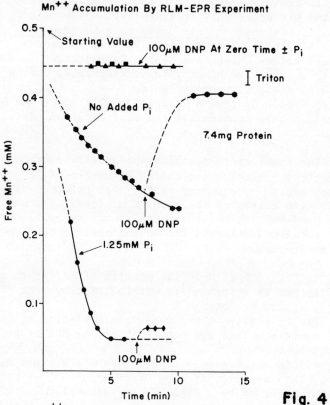

Fig. 4

The uptake of Mn^{++} by liver mitochondria followed by EPR. The Mn^{++} spectrum was scanned continuously and the amount of free Mn^{++} at any instant determined from the height of the peaks. Conditions as for Fig. 2.

---

for these sites.

(2) During Mn^{++} accumulation by the mitochondria the enhancement fell. In the absence of P_i, the value tended towards 0.5. This would occur if the Mn^{++} was bound to some component in which all the water ligands had been replaced or bound at a site which was unaccessible to water in the bulk of the medium. It is consistent with binding to phospholipid, the hydrophobic nature of which, might be expected to reduce the accessibility of water, or, it is consistent with binding in a chelated form of the type of $Mn^{++}EDTA$. It seems unlikely that protein binding is involved since this process is more usually associated with enhancement values much greater than

unity, although more rarely, enhancement <1 is observed if the ion is tightly bound within the protein molecule.

(3) In the presence of P_i, Mn^{++} was bound by mitochondria in a form which had a very low enhancement value, tending towards 0.1. In this instance, it is suggested that this low value was observed because the Mn^{++} was out of solution, possibly as $Mn_3(PO_4)_2$, and that the access of water was limited by this.

TABLE III
Components of isolated rat liver mitochondria

Phosphate compounds	(μmole/gm. protein)
Total P	690
Acid soluble	140 (P_i, 30-40; adenine nucleotides, 14-19)
Phospholipid	445
Pentose nucleic acid	74
Amino acids	10,000
Mn^{++} bound in absence of P_i	200-300

Table III lists the quantities of some of the components of isolated liver mitochondria. It can be seen that there is too little acid - soluble phosphate [and ATP would give an ϵ_b of approximately 1.5 (Cohn, 1963)] and pentose nucleic acid (which would give a high ϵ_b value) to account for the binding of Mn^{++}, in the absence of added P_i. Only phospholipid and protein occur in sufficient amounts. The latter seems unlikely to be responsible for the binding of Mn^{++} since enhancement is usually observed when Mn^{++} is bound to protein. The most likely candidate appears to be phospholipid. Slater and Cleland (1953) have suggested that Ca^{++} is bound to the phospholipid of heart mitochondria.

Fig. 5 is a schematic representation of what is considered, at the present time, to be the course of events during Mn^{++} accumulation.

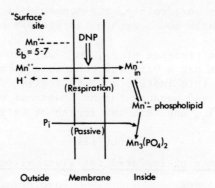

Fig. 5. Proposed scheme for Mn^{++} accumulation by mitochondria.

Table IV summarises some of the findings discussed above.

TABLE IV

Mn^{++} binding and accumulation by isolated rat liver mitochondria

Type	Amount (μatom/gm. protein)	Enhancement	Remarks
"Surface" binding	ca. 30	5-7	Occurs rapidly, insensitive to DNP, antimycin, rotenone. Mg^{++}, Ca^{++} compete.
Respiration-dependent accumulation (no P_i)	200-300	less than 0.5	Both inhibited and reversed by DNP and respiratory inhibitors. Possibly bound to phospholipid.
Respiration-dependent accumulation, P_i present	Greater than 2000	Possibly zero	Inhibited but not reversed by DNP and respiratory inhibitors. Greater part of Mn^{++} precipitated as $Mn_3(PO_4)_2$.

It has been claimed that ATP is required for the respiration-dependent accumulation of Ca^{++} by mitochondria. (Brierley et al. 1962; Lehninger et al. 1963a). The probable reason for this requirement is that mitochondria undergo extensive swelling in the presence of P_i and Ca^{++}. In the experiment shown in Fig. 6, on addition of 1 μmole Ca^{++}, the mitochondria

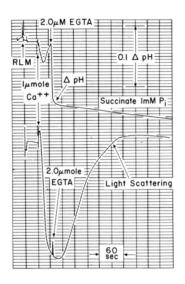

Fig. 6. pH - and light-scattering changes on addition of Ca^{++} to mitochondria. Conditions as for Fig. 2 with the further addition of 1 mM P_i; 5 mM succinate was the substrate, instead of glutamate-malate.

---

underwent an apparent contraction (upward deflection on lower trace) followed by extensive swelling (downward deflection). Simultaneous measurement of the changes in pH (upper trace) indicated an acid production followed by return to the initial pH value, the latter possibly associated with increased permeability of the mitochondrial membrane, due to swelling and release of accumulated Ca^{++} and P_i. In other experiments the presence of ATP prevented swelling; it is possible that this occurred because the free Ca^{++} concentration was held low by the chelating action of ATP.

When P_i was replaced by arsenate, even with ATP present, rapid swelling occurred. This would account for the failure to observe Ca^{++} accumulation when arsenate was present.

On addition of EGTA to Ca^{++}-P_i swollen mitochondria, after a short lag, the mitochondria contracted, in a respiration-dependent, DNP-inhibited, oligomycin-insensitive process (Chappell and Greville, 1963; see also Lynn et al. 1963). In the presence of antimycin 2 mM-ATP caused reversal of swelling; in this case oligomycin inhibited.

I would like to thank Professor Britton Chance for his hospitality and encouragement of members of the Johnson Foundation, particularly Drs. Estabrook and Mildvan, for many stimulating discussion.

REFERENCES

1. Brierley, G.P., Backmann, E. and Green, D.E., Proc. Nat. Acad. Sci., 48, 1928 (1962).
2. Brierley, G.P., Murer, E., Backmann, E. and Green, D.E., Fed. Proc., 22, 526 (1963).
3. Chance, B. in Wolstenholme, G.E.W. and O'Connor, C.M. (Editors) Ciba symposium on regulation of cell metabolism, J. and A. Churchill Ltd., London, 1959, p. 91.
4. Chappell, J.B., Biochem. J., 84, 62p(1962).
5. Chappell, J.B. and Greville, G.D., Fed. Proc., 22, 526 (1963).
6. Chappell, J.B., Greville, G.D. and Bicknell, K.E., Biochem J., 84, 61p(1962).
7. Cohn, M., Biochemistry, 2, (1963), in press.
8. Fiske, C.H. and SubbaRow, Y., J. Biol. Chem., 66, 375 (192
9. King, T.E. and Davidson, L., J. Chem. Phys., 29, 787 (1958
10. Lehninger, A.L., Rossi, C.S. and Greenawalt, J.W., Biophys. Biochem. Res. Comm., 6, 444 (1963a).
11. Lenhinger, A.L., Rossi, C.S. and Greenawalt, J.W., Fed. Proc., 22, 526 (1963b).

12. Lynn, W.S., Brown, R.H., Fortney, S. and Clancy, T., Fed. Proc., 22, 526 (1963).
13. Slater, E.C. and Cleland, K.W., Biochem. J., 55, 566 (1953).
14. Vasington, F.D. and Murphy, J.V., J. Biol. Chem., 237, 2670 (1962).

DISCUSSION

<u>Klingenberg</u>: I just want to make sure where the binding action of the manganese takes place. I propose that the transporting sites are on the inner membranes and not the outer membranes; transport occurs mainly inside the mitochondria.

<u>Chappell</u>: I wouldn't care to be too specific at this stage.

<u>Klingenberg</u>: What structure is "inside," here?

<u>Chappell</u>: This could be an intercristal space, but I wouldn't be too specific about this. It would seem the natural thing to assume the outer membrane is the site of transport, but I see no real reason for believing it should be there.

<u>Green</u>: In all the studies which Dr. Brierley and I have been carrying out on the accumulation of Mg^{++} and Ca^{++} by mitochondria, we have found an invariant association between the accumulation of Mg^{++} or Ca^{++} and the accumulation of phosphate. The molar ratio of the cation:anion lies between 1.5 and 1.8. I find it, therefore, of great interest that you are able to observe accumulation of Mn^{++} in the absence of added phosphate. I would like to throw out a possible explanation for this discrepancy. From your data the accumulation process for Mn^{++} that is operative when phosphate is not added appears to be indistinguishable from the process operative in the presence of added phosphate.

<u>Chappell</u>: No, there is one very important difference, and that is that the effect in the absence of phosphate is reversible by switching off energy-coupled metabolism, e.g. by adding an uncoupling agent.

<u>Green</u>: Suppose that you have a mixed population of mitochondria, some of which are intact while the rest are defective with respect to the membrane. Phosphate from the interior of the latter mitochondria would trickle out and become available for accumulation in association with Mn^{++} by the intact mitochondria. The levels of Mn^{++} that you are working with are relatively small. Thus the amount of endogenous phosphate in the mitochondria might be sufficient to provide the requisite amounts of phosphate required for the accumulation. When you have accumulated Mn^{++} in the absence of phosphate, what is the ratio of Mn^{++} to phosphate in these mitochondria?

Chappell: It is extremely low. It is a matter of about 200 to 300 manganese to 20 or 30 phosphates; i.e., about 10 Manganese atoms for each phosphate.

Racker: I asked Dr. Chappell what the difference is, if any, when phosphate is present.

Chappell: It is a matter of either 10 manganese per phosphate without added phosphate, and 1.5 manganese per phosphate when phosphate is added. So there is a difference of an order of magnitude.

Cohn: Also, the enhancement factor of the bound manganese is different when phosphate is added. Therefore, the manganese must be bound to something else in this case. This is "per manganese bound."

Racker: Dr. Green, can you tell us how you explain the difference between your experiments and Dr. Chappell's in the response with dinitrophenol?

Green: I am assuming that there would be a difference in the stability of the mitochondria under these conditions. Suppose that mitochondria in presence of Mn^{++} but in absence of phosphate were fragile. Such mitochondria exposed to dinitrophenol would leak accumulated Mn^{++} more readily than mitochondria that had accumulated Mn^{++} in presence of added phosphate. Dr. Chappell had indeed mentioned that the mitochondria showed a tendency to disintegrate.

Chappell: It is very important to know the concentration of free Mn^{++} inside the mitochondria, before we can talk of an active transport system. We must know not only the concentration of free Mn^{++}, but also the membrane potential, since an ion is actively transported only when it is moved against an electrochemical gradient. The electrochemical activity of an ion (μe) is given by:-

$$\mu_e = \mu_c + Z \cdot \psi$$

where μ_c, the chemical activity is given by RTlog (activity of the ion), Z is the valence and ψ the electrical potential. If the greater part of the Mn^{++} were bound inside the mitochondrion or if the membrane potential favored Mn^{++} accumulation, i.e., the inside of the mitochondrion were negative with respect to the outside, then the uptake of Mn^{++} could be passive, but only occur when coupled respiration occurred, rather than being dependent on coupled respiration in an energetic sense. Dramatic changes in the swelling and shrinking of mitochondria occur when electron transport is instituted or inhibited, and these volume changes presumably reflect changes in the permeability of the mitochondrial membrane associated with changes in the level of activity of the respiratory chain (1). The accumulation of divalent cations by mitochondria may well reflect an active transport process, but a better choice of terms at our present level of knowledge would be "respiration-dependent accumulation."

Racker: It seems to me that very critical experiments could be performed with submitochondrial particles on the EPR apparatus, if you can detect a respiration-dependent manganese binding by the membrane itself. My question is whether you have done similar experiments with phosphorylating submitochondrial particles?

Chappell: Dr. Hommes has done this with beef heart mitochondria and has observed binding that was respiration independent

Hommes: It is completely independent of the respiration.

Packer: And these were actively phosphorylating particles?

Hommes: Yes.

Packer: This is contrary to the ability of submitochondrial particles to phosphorylate and to manifest mechano-chemical changes.

Chappell: One ought to test for the binding of potassium in a respiration-dependent system.

Klingenberg: I wish to report an experiment on manganese and magnesium interaction. If you observe DPNH in a mitochondrial suspension and pretreat with magnesium, nothing happens, but if you then add manganese, the cycle is much longer and the respiration much slower than in a control experiment without magnesium. This means that in the presence of magnesium the apparent P:O is lower. I would suggest that the magnesium which may be bound is released and transported inward during the cyclic transition from State 3 to State 4. This is possibly an important physiological pehnomenon because the P:O is lower in the presence of magnesium than in the absence of magnesium under incubation conditions.

Chappell: What was the concentration of magnesium?

Klingenberg: This is an excess of magnesium.

Chappell: I am surprised that the pyridine nucleotide becomes reduced again, because we find a very marked correlation of pyridine nucleotide cycles with the disappearance of the ion into the mitochondrion. Do you mean P:O, or the length of the cycle?

Klingenberg: Both. I have measured the respiration and calculated the P:O from the respiratory control.

Löw: It seems to me that you have very convincingly demonstrated *in vitro* a mechanism that was suggested to exist *in vivo* some years ago (2). You suggest that the manganese ions are trapped inside the mitochondrion as insoluble manganese phosphate. Can you suggest any mechanism with which the manganese can be transported <u>out</u> again?

Chappell: I cannot, but the mitochondria accumulate it with remarkable efficiency.

REFERENCES

1. Chappell, J. B., and Greville, G. D., Biochem. Soc. Symp. No. 23, p. 39 (1963).
2. Maynard, L. S., and Cotzias, G. C., J. Biol. Chem., <u>214</u>, 489 (1955).

ION ACCUMULATION IN HEART MITOCHONDRIA

G.P. Brierley

Institute for Enzyme Research, University of Wisconsin
Madison, Wisconsin

Recent studies from several laboratories (1-8) have made it abundantly clear that mitochondria possess the ability to accumulate relatively large amounts of certain divalent cations and inorganic phosphate (P_i). In the studies presented here the accumulation of Mg^{++} and P_i is compared with that of Ca^{++} and P_i by isolated beef heart mitochondria.

TABLE I

The Accumulation of Magnesium and Phosphate by Beef Heart Mitochondria

	Mg^{++}	P_i
	(mμmoles/mg	of protein)
Untreated mitochondria	50	30
No substrate added	75	10
Complete system (succinate)	1800	1000
Complete system (pyruvate + malate)	1650	900
Complete + Antimycin (0.2 μg/mg)	75	40

The complete system consisted of 5 mg of mitochondrial protein incubated for 15 min at 38° in 3 ml of 0.25 M sucrose containing 17 mM Mg^{++}, 3.3 mM potassium phosphate, 3.3 mM succinate or pyruvate and malate, and 10 mM Tris (pH 7.4). The mitochondria were separated from the medium by centrifugation and Mg^{++} and P_i were determined as previously described (1, 2).

These studies were performed in collaboration with Dr. D.E. Green, E. Bachmann, E. Murer, and D.G. Hadley of the Institute for Enzyme Research.

This work was supported in part by National Heart Institute research grant H-458 (USPH) and Atomic Energy Commission Contract AT (11-1)-909. Meat by-products were generously supplied by Oscar Mayer and Company, Madison.

TABLE II

The Efficiency of Mg^{++} and P_i Accumulation

Substrate	Rate		Ratio
	P_i	O_2	P_i/O
Succinate (3.3 mM)	0.06	0.08	0.75
Pyruvate + malate (3.3 mM each)	0.08	0.07	1.14
Pyruvate + malate (0.7 mM each)	0.105	0.055	1.91

P_i rates are μmoles of P_i accumulated/min/mg of protein in a 3 to 4 min incubation. O_2 rates are μatoms of O_2 consumed/min /mg of protein in the same period.

Mitochondria of beef heart muscle accumulate up to 1.8 μmoles of Mg^{++} and 1.0 μmole of P_i when incubated aerobically with substrate (Table I). No accumulation occurs in the absence of substrate or when respiration is inhibited by Antimycin, cyanide, or other inhibitors of electron transport. Magnesium and phosphate accumulation can be supported by a number of substrates (1). Although the kinetics of the accumulation vary with the substrate employed, in every case it is clear that (a) the two ions always accompany each other and (b) the ratio of Mg^{++} to P_i accumulated remains relatively constant and the value of this ratio averages 1.8. The uptake of these ions occurs at rather low rates of respiration (Table II), but the process can be quite efficient. The ratio [μmoles of P_i transported] to [μatoms of O_2 consumed] can approach a value of two.

The pH of the incubation medium decreases during the accumulation of Mg^{++} and P_i. There is close to a one to one correspondence between P_i bound and H^+ released and the requirements for H^+ release are identical with those for Mg^{++} and P uptake. When isolated mitochondria, which have accumulated large amounts of Mg^{++} and P_i, are suspended in water the suspension shows a rather alkaline pH (9.0-9.5). The H^+ released does not represent a net increase in acid in the system since disruption of the mitochondrial membranes by the addition of Triton X-100 causes the pH of the suspending medium to return to its original value. We have suggested (1, 2) that the exaggerated accumulation of Mg^{++} and P_i results from a precipitation of alkaline Mg^{++} phosphate salts within the mitochondrion and that the precipitation of these salts from solutions of neutral pH could explain the extensive release of H^+.

ENERGY-LINKED FUNCTIONS OF MITOCHONDRIA

The accumulation process is closely related to the energy transfer system. The following uncouplers of oxidative phosphorylation all give half-maximal inhibition of ion accumulation and of oxidative phosphorylation at virtually the same concentration: Gramicidin, arsenate, dinitrophenol, dicoumarol, Ca^{++}, azide, and m-Cl-carbonyl cyanide phenylhydrazone. The actual synthesis of ATP does not appear to be required for the accumulation, however, since the process is completely insensitive to oligomycin at concentrations far in excess of those necessary to abolish ATP synthesis (1-3).

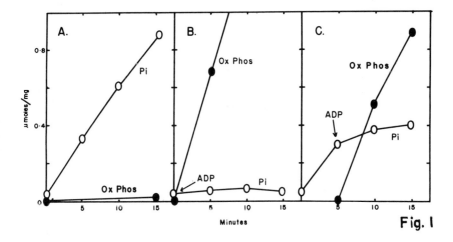

Fig. 1

The apparent competition between the accumulation of P_i (and Mg^{++}) and oxidative phosphorylation. A. No added ADP. The open circles show the accumulation of P_i as a function of time at 38° in a medium of 0.25 M sucrose, 3.3 mM succinate, 17 mM $MgCl$, 3.3 mM potassium phosphate (labeled with P^{32}), and 10 mM Tris (pH 7.4). The closed circles show the incorporation of P^{32} into organic phosphate in the same reaction (abbreviated "ox phos"). B. The identical reaction medium as shown in part A but with ADP (3.3 mM) plus hexokinase and glucose added at time zero. C. Identical with part B except that the ADP plus hexokinase and glucose were added 5 min after the incubation had begun. In each case samples were withdrawn at the indicated time; the intramitochondrial P_i was determined as previously described (1, 2) and the total P^{32} in organic phosphorus was determined by the method of Lindberg and Ernster (10).

The accumulated ions can diffuse back into the medium when incubation is prolonged (2) and this process is accelerated when respiration is inhibited by lowering of the temperature or by the addition of Antimycin, or by the addition of an uncoupler of oxidative phosphorylation.

The experiments presented in Fig. 1 show that oxidative phosphorylation and the uptake of Mg^{++} and P_i appear to be competitive reactions. In the absence of added ADP there is negligible oxidative phosphorylation and the accumulation of Mg^{++} phosphate proceeds to high levels. When ADP (in presence of glucose + hexokinase) is added at the start of the incubation, there is negligible ion accumulation while oxidative phosphorylation proceeds at the expected high rate. If ADP and hexokinase are added after 5 min of incubation, the rate of P_i accumulation decreases immediately and a high rate of oxidative phosphorylation is established. This condition does not completely eliminate P_i accumulation, and it should also be noted that during the vigorous phosphorylation reaction the high intramitochondrial level of P_i established during the first 5 min of incubation is not depleted.

TABLE III

Effect of Adenine Nucleotides on the Accumulation of P_i

Nucleotide Added	μmoles P_i/mg protein	
	no oligomycin	+ oligomycin (1 μg/mg)
None	0.80	0.80
AMP	0.15	0.45
ADP	0.25	0.50
ATP + hexokinase	0.03	0.44

The substrate was succinate (3.3 mM) and the nucleotide concentration was 3.3 mM. The experiments were carried out as described in the legend for Table I.

The inhibition of Mg^{++} phosphate accumulation by adenine nucleotides is shown in Table III. It is apparent that AMP and ADP added at the start of the reaction markedly inhibit the accumulation reaction. This inhibition is even more striking in the presence of hexokinase. Inhibition of ion accumulation by adenine nucleotides can be removed to a large extent by the addition of oligomycin to block the synthesis of ATP.

In the absence of respiration a limited accumulation of

Mg^{++} and P_i can be supported by ATP. This reaction differs from the substrate-supported reaction in that it is inhibited by the same concentrations of oligomycin which inhibit oxidative phosphorylation. The reaction is also inhibited by uncouplers of oxidative phosphorylation.

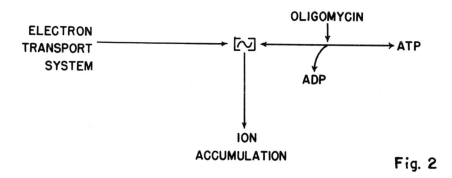

Fig. 2

The relationship of the postulated high-energy intermediate [∿] to ion accumulation and to ATP synthesis by oxidative phosphorylation.

These considerations permit us to postulate the simplified reaction sequence represented in Fig. 2. The features of this scheme can be summarized as follows:

(a) Electron transport generates one or more high-energy intermediates of oxidative phosphorylation which can support the accumulation of Mg^{++} and P_i without the involvement of the oligomycin-sensitive site.

(b) In the absence of respiration ATP can support accumulation by an oligomycin-sensitive reaction. This reaction could possibly represent a reversal of the normal pathway for ATP synthesis by oxidative phosphorylation.

(c) ADP inhibits ion accumulation by stimulating oxidative phosphorylation which competes with ion accumulation for the high-energy intermediates; inhibition of oxidative phosphorylation by oligomycin removes much of this inhibition.

(d) While it is not yet certain whether the Mg^{++} or the P_i (or both) are transported into the mitochondrion by the energy-requiring process, it appears that the exaggerated accumulation of these ions results from the precipitation of Mg^{++} phosphate within the mitochondrion, and such a precipitation could explain the extensive release of H^+.

Beef heart mitochondria also accumulate Ca^{++} and P_i (3)

and the uptake of these ions has several features in common with the Mg^{++} reactions just described. The two ions are accumulated together under all conditions tested and the ratio of Ca^{++} to phosphate remains relatively constant at about 1.6. Hydrogen ion is released and ATP, when present, is hydrolyzed. Both the total amount of accumulation and the rate of accumulation of Ca^{++} are greater than the corresponding values for Mg^{++} uptake. Electron micrographs prepared by Dr. D. B. Slautterback reveal large electron dense deposits within the mitochondria following the accumulation of Ca^{++} phosphate. The amount of electron-dense material is proportional to the amount of Ca^{++} and P_i accumulated.

TABLE IV

Ca^{++} Accumulation Supported by ATP

	Ca^{++} uptake (μmoles/mg protein)	ATPase (μmoles/mg protein)
Fresh Mitochondria		
No addition	1.44	1.59
Oligomycin (1 μg/mg)	0.23	0.40
Azide (10^{-4} M)	0.23	0.19
Aged Mitochondria		
No addition	0.11	0.17

Mitochondria (2.5 mg) of protein were incubated for 3 min at 38° in a medium of 0.25 M sucrose and 3 mM imidazol (pH 7.0) containing 3.3 mM ATP, 2.3 mM Ca^{++} labeled with Ca^{45}, 10 mM Mg^{++}, 3 mM P_i, and 0.2 μg of Antimycin per mg of protein. Ca^{++} uptake was determined by the increase in radioactivity of the isolated mitochondria and ATPase by the increase in total P_i following incubation.

At high concentrations of Ca^{++} (2-3 mM), where oxidative phosphorylation is completely uncoupled, two reactions pathways can readily be distinguished. The most obvious reaction is supported by ATP (Table IV). This ATP-supported accumulation occurs in the absence of respiration and is accompanied by a Ca^{++} dependent ATPase. The loss of the Ca^{++} dependent ATPase always parallels the loss of ability to accumulate Ca^{++} and phosphate by this pathway. This reaction is inhibited by oligomycin and low concentrations of azide, and it also disappears in mitochondria which have been aged in the absence of substrate. The ATP-dependent accumulation of Ca^{++} requires Mg^{++}, but does not require added phosphate; it is inhibited by uncouplers of oxidative phosphorylation but usually at concentrations higher than those necessary to uncouple phosphorylation. Ca^{++} accumulation in long term incubations does not proceed to high levels in the absence of added ATP. However, in the presence of substrate as well as ATP, the rate of Ca^{++} accumulation is higher than in the absence of substrate, and a portion of this reaction is inhibited by inhibitors of respiration and insensitive to oligomycin (3).

TABLE V

Ca^{++} Accumulation Supported by Substrate

Exp.	Conditions	Ca^{++} Uptake (μmoles/mg protein)		
		No addition	Oligomycin (1 μg/mg)	Antimycin (0.2 μg/mg)
1.	Succinate (3 mM), ATP (3mM) Oligomycin (1 μg/mg)-3 min at 38°	--	1.25	0.15
2.	Aged mitochondria, succinate (3 mM) ATP (3 mM)-3 min at 38°	1.38	1.00	0.15
3.	No ATP, succinate (3 mM) 20 sec at 38°	0.40	0.39	0.03
4.	No ATP, succinate (3 mM) Ca^{++} (0.3 mM) - 2 min at 38°	0.20	0.18	0.01

All experiments were incubated at pH 7.0 in 0.25 M sucrose containing 3 mM imidazol, 3 mM phosphate, and 10 mM $MgCl_2$. Exps. 1-3 contained 2.3 mM Ca^{++}. Mitochondria designated "aged" were shaken for 30 min at 38° in a medium of 0.25 M sucrose containing 0.01 M Tris (pH 7.5).

This substrate-supported portion of the reaction can be studied by the four different methods listed below (and in Table V) and shows almost the identical requirements as the uptake of Ca^{++} by digitonin particles reported by Vasington (5).
1. The substrate-supported reaction can be studied in isolation by the addition of oligomycin and low concentrations of ATP. In this case the ATP-supported reaction is minimized and the accumulation of Ca^{++} becomes completely sensitive to Antimycin and cyanide. Even in the presence of oligomycin,

ATP addition is essential in order to obtain appreciable accumulation of Ca^{++}. It is well known that high concentrations of Ca^{++} are destructive to mitochondria and that ATP counteracts this effect. It is, therefore, possible that under these conditions there is a secondary requirement for ATP in order to maintain the integrity of the mitochondrion.

2. As has been mentioned, aged mitochondria lose capacit to carry out the ATP-supported Ca^{++} accumulation reaction; these mitochondria still require ATP for the accumulation, bu the accumulation of Ca^{++} is now relatively insensitive to oli gomycin and is almost completely inhibited by Antimycin.

3. In absence of ATP but in presence of substrate there is an accumulation of Ca^{++} and phosphate which can be observe in the first few seconds of incubation. The accumulated ions diffuse back into the medium after about one minute unless AT is also added. This uptake in the absence of ATP requires substrate, inorganic phosphate, and usually requires magnesiu it is insensitive to oligomycin but is inhibited by cyanide, Antimycin, dinitrophenol and other uncouplers.

4. When low concentrations of Ca^{++} (0.3 mM) are present during incubation the oligomycin-insensitive, substrate-dependent reaction permits accumulation of almost all of the avail able Ca^{++} without the necessity for added ATP. Again, this reaction is inhibited by Antimycin and cyanide, requires inor ganic phosphate, and is extremely sensitive to dinitrophenol and the other uncouplers of oxidative phosphorylation.

These studies indicate that Ca^{++} phosphate accumulation like Mg^{++} phosphate accumulation can be supported by either a oligomycin-insensitive, substrate-dependent, pathway or by an ATP-dependent, oligomycin-sensitive, pathway. When high concentrations of Ca^{++} are present, an additional requirement for added ATP, apparently to protect the mitochondria from the des tructive effects of high levels of Ca^{++}, becomes apparent.

It can be concluded that Ca^{++} accumulation behaves much like Mg^{++} accumulation in that both can be supported either b substrate in the absence of ATP synthesis or by ATP in the ab sence of respiration. Ca^{++} inhibits Mg^{++} accumulation in tha neither ion accumulates in the absence of ATP (probably due t the extensive deterioration of the mitochondria); in the presence of ATP, both ions accumulate but the amount of Ca^{++} boun is large compared to the amount of Mg^{++}. Ca^{++} accumulation i a rapid reaction and is accompanied by high rates of respiration typical of uncoupled systems and by high rates of hydrolysis of ATP; Mg^{++} accumulation is much slower and occurs at rather low rates of oxidation and low rates of hydrolysis of

ATP. At present it is not possible to state whether Ca^{++} and Mg^{++}-phosphates are accumulated by identical mechanisms, but the studies presented here show that the energy-transfer requirements of the two reactions are quite similar.

REFERENCES

1. Brierley, G.P., Bachmann, E. and Green, D.E., Proc. Nat'l. Acad. Sci., U.S., 48, 1928 (1962).
2. Brierley, G.P., Murer, E., Bachmann, E. and Green, D.E., (in preparation).
3. Brierley, G.P., Murer, E. and Green, D.E., Science, 140, 60 (1963).
4. Vasington, F.D. and Murphy, J.V., J. Biol. Chem., 237, 2670 (1962).
5. Vasington, F.D., Fed. Proc., 22, 474 (1963).
6. Chappell, J.B., Greville, G.D. and Bicknell, K.E., Biochem J., 84, 61p (1962).
7. Lehninger, A.L., Rossi, C.S., and Greenawalt, J., Biochem. Biophys. Res. Comm., 10, 444 (1963).
8. DeLuca, H.F. and Engstrom, G.W., Proc. Natl. Acad. Sci., U. S., 47, 1744 (1961).
9. Slater, E.C. and Cleland, K.W., Biochem. J., 55, 566 (1953).
10. Lindberg, O. and Ernster, L., in Glick, D. (Editor) Methods of Biochemical Analyses, III, Academic Press, Inc., N.Y., 1956, p. 1.

DISCUSSION

Klingenberg: I wonder whether there are differences between the liver and the heart mitochondria? Did you study liver mitochondria?

Brierley: No, we haven't.

Chappell: We have studied pigeon heart mitochondria; they behave in a somewhat similar fashion.

Klingenberg: I think that liver mitochondria would not accumulate such a large amount of Mg^{++}.

Brierley: Recent studies by Sallis et al. (1) have shown that liver mitochondria accumulate massive amounts of Mg^{++} and P_i, but that the process requires parathyroid hormone. We have not yet tested heart mitochondria under their conditions.

Klingenberg: I have some data which bear on this point (Fig. 1). If oxidation of pyridine nucleotide and activation of respiration are an indication of metal ion accumulation, then Mg^{++} has a very short-lived effect, although it is in excess and there is phosphate present. In that case, Ca^{++} gives an extensive oxidation, which is followed by an inhibition of respiration.

Brierley: All of the Mg^{++} values which I have reported were obtained by chemical analysis. Since Mg^{++} accumulation occurs at low rates of respiration, activation of respiration may not be a reliable criterion for this reaction.

Pressman: The data on this slide clearly show that magnesium does accumulate in rat liver mitochondria in response to relatively high extramitochondrial magnesium concentrations. It is my understanding that the basic technique Dr. Brierley used to separate the mitochondrially bound magnesium from that in the medium was essentially the same as ours (2). Figure 2 also indicates that the binding of magnesium and guanidine are to some extent mutually competitive. The level of bound magnesium rises progressively to a saturation point as the extramitochondrial magnesium concentration increases, but at all concentrations, some bound magnesium is diminished by 20 mM guanidine. Conversely, the amount of added guanidine which

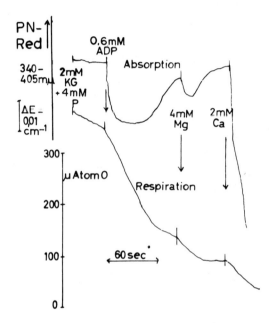

Fig. 1 (Klingenberg). The influence of ion transport on the respiratory chain. Comparison of the effects of Mg^{++}, Ca^{++}, and ADP on the redox state of pyridine nucleotides and respiration in liver mitochondria. Oxidation of pyridine nucleotides is accompanied by an increased respiration also with Mg^{++} and Ca^{++}. Simultaneous recordings of absorption and oxygen concentration in a mitochondrial suspension.

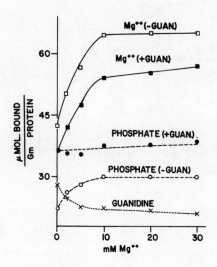

Fig. 2. (Pressman). Effect of Extramitochondrial Mg^{++} on the Binding of Mg^{++}, Phosphate, and Guanidine. The medium consisted of: phosphate, 4 mM; $MgCl_2$, 6 mM; KCl, 30 mM; ATP, 0.4 mM; sucrose, 220 mM. Samples were incubated 4 min, 30° before chilling and separation of mitochondria from the medium.

accumulates in the mitochondria is progressively diminished by increasing concentrations of extramitochondrial magnesium. Other data, indicating that guanidine accumulation requires substrate derived energy but is not inhibited by oligomycin, also tends to suggest a close relationship between the accumulation by mitochondria of guanidine and divalent metal cations. Guanidine accumulation appears to be a general property of mitochondria as we have observed it in preparations from pigeon heart and rabbit kidney as well as rat liver. Despite the fact that guanidine forms no insoluble phosphate and has only one-half the charge of the divalent metal ions, accumulation of guanidine is accompanied by the simultaneous accumulation of from 2/3 to one molar equivalents of phosphate, approximately the same ratio as obtained with divalent cations. This suggests that although the stoichiometry of cation:phosphate binding is consistant with the formation of an $X_3(PO_4)_2$ precipitate, it may actually be determined by some alternative characteristic of the binding process. Clearly, this must be so in the case of the singly charged guanidine. Another feature of guanidine accumulation reminiscent particularly of manganese accumulation is that it does occur in the absence of added phosphate, but to a diminished extent.

Jöbsis: We have been studying the inhibitory effect of oligomycin on the so-called sodium and potassium activated ATP-ase from membranes and we find that it inhibits this reaction at a level considerably higher than it inhibits oxidative phosphorylation. So, my question is: at what levels would you get your oligomycin inhibition?

Brierley: The accumulation of Ca^{++} and Mg^{++} which is supported by ATP is inhibited by oligomycin at the same concentration that inhibits oxidative phosphorylation.

Jöbsis: Does anyone dare to make an estimate as to how much Ca^{++} would be in the cytoplasm and how much in the mitochondria under normal conditions?

Chance: We have data on the Ca^{++} affinity; perhaps someone else has the amount of Ca^{++} that is free in the cytoplasm.

Brierley: The heart mitochondria as isolated contain on the order of 30 µmoles of Ca^{++} per mg protein. However, the early studies of Slater and Cleland (3) have shown that this value may have little relation to the actual concentration of Ca^{++} in the mitochondria *in situ*.

Chance: Dr. Jöbsis wishes to know how much external Ca^{++}

could be free, not how much is inside. Thus the K_m for Ca^{++} is important.

Brierley: The K_m for Ca^{++} is about 1 mM as measured under our conditions of ATP, Mg^{++}, phosphate, and substrate.

Chappell: This value has, of course, to be corrected for the binding of Ca^{++} by the ATP you have present in your system. The value of the apparent K_m for free Ca^{++} would be much less than the one you gave.

Chance: A K_m of 1 mM is difficult to understand. It seems to be rather different from what Dr. Chappell found and what we have found for mitochondria. The K_m should be small because these reactions are "driven" so hard by Ca^{++} phosphate precipitation that the energy balance is very much in favor of the insoluble form. Is this a possibility?

Klingenberg: Yes, this is possible, but the question is whether this is an active process, whether the phenomena you observed are energy-transfer-dependent, and to what degree.

Brierley: The accumulation of both Mg^{++} and Ca^{++}-phosphates is inhibited by uncouplers of oxidative phosphorylation and by inhibitors of electron transport at the concentrations which would be expected for an energy-dependent process. The ion accumulation which is dependent on substrate is not sensitive to oligomycin, however, and for this reason we have suggested that the process is energized by an intermediate of oxidative phosphorylation.

Chance: It is possible that there are conditions under which the mitochondria will retain Ca^{++} phosphate and others under which they will not. This is perhaps the more important criterion when you are forming an insoluble precipitate.

Chappell: To this point, if one adds low levels of Ca^{++} to mitochondria one stimulates respiration and the pyridine nucleotides become oxidized. The mitochondria are in a "pseudo-State 3" and remain in this condition until either phosphate or a metal-chelating agent are added. When phosphate is added respiration is stimulated even more and pyridine nucleotide goes even more oxidized but then, when the Ca^{++} has been taken up, the mitochondria go into State 4. It appears that, in the absence of added phosphate, the accumulated divalent metal ions are held in the mitochondria only while energy is provided by coupled respiration. In the presence of phosphate, once accumulation has occurred, energy is no longer required. However, addition of dinitrophenol or respiratory inhibitors does cause release of Ca^{++} and phosphate, but this is because extensive swelling of the mitochondria occurs. It doesn't happen when Ca^{++} is replaced by Mn^{++}.

At low levels of Ca^{++}, insufficient to cause swelling, nearly all the Ca^{++} is firmly bound to the mitochondria, even in the absence of phosphate.

Slater: I would like to confirm that. In our old experiments rat heart sarcosomes prepared in the presence of EDTA and then washed at the centrifuge to remove the EDTA, took up all the Ca^{++} when suspended in a 0.1 mM solution of $CaCl_2$. Sarcosomes prepared in the absence of EDTA took up half the Ca^{++} when suspended in 0.1 mM $CaCl_2$, the Ca^{++} content of the sarcosomes increasing from 83 μmoles/g protein to 127 μmoles/g protein.

REFERENCES

1. Sallis, J. D., et al., Biochem. Biophys. Res. Comm., 10, 266 (1963).

2. Pressman, B. C., J. Biol. Chem., 232, 967 (1958).

3. Slater, E. C., and Cleland, K. W., Biochem. J., 55, 566 (1953).

CALCIUM-STIMULATED RESPIRATION IN MITOCHONDRIA

Britton Chance

Johnson Research Foundation, University of Pennsylvania
Philadelphia 4, Pennsylvania

It appears to be highly appropriate to conclude a discussion on energy-linked functions of mitochondria with studies of their reactions with divalent cations. Cation accumulation and the energy-linked reduction of DPN do not depend on the direct mediation of ATP but instead represent reactions driven by the splitting of high energy precursors of ATP in a reaction of the general type

$$X \sim I \rightleftarrows X + I \qquad (1)$$

In one case, DPN is reduced; in the other, divalent cations are accumulated by the mitochondria.

In 1953 Siekevitz and Potter (1) and Potter et al (2) observed a three-fold respiratory stimulation in rat liver mitochondria in the presence of 500 µM Ca^{++}, and compared its action with that of DNP. Also in 1953, Slater and Cleland (3) measured the binding of Ca^{++} by heart sarcosomes.

Our 1955 experiment, reproduced in Fig. 1, illustrates important properties of Ca^{++} binding and respiratory activation (4). In the control assay, ADP addition caused respiratory stimulation to 1.0 µM/sec and a cycle of oxidation and reduction of pyridine nucleotide. Respiratory activity was measured polarographically with the vibrating electrode and pyridine nucleotide was measured spectrophotometrically with 340 mµ as a measuring wavelength and 374 mµ as a reference wavelength (6, 7). (A decrease of oxygen is represented as an upward deflection of the trace in contrast to the usual convention (5).)

A second addition consisted of calcium at a concentration of approximately 3 times that of the previously added ADP. Respiration was immediately stimulated to an even faster rate, 170 % of the rate obtained with ADP. Also the stimulation of respiration was accompanied by an oxidation of reduced pyridine nucleotide; after 26 µM oxygen was consumed respiration slowed and pyridine nucleotide became more reduced. The effects of calcium simulate those of ADP as an activator of the phosphorylation reaction, i.e., as an energy acceptor. It is notable that the reaction with Ca^{++} differs from that of uncoupling agents in that the effects of low Ca^{++} are reversible; those of uncoupling agents such as DNP are not. In accordance with the results of previous workers, addition of higher concentrations of calcium (400 μM) destroys the mitochondria, causing the pyridine nucleotide to become highly oxidized and respiration to start rapidly but to fall to an inhibited rate of 0.8 μM/sec.

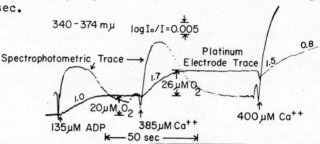

Fig. 1. Activation of respiration and oxidation of DPNH during accumulation of calcium by guinea pig liver mitochondria: a comparison with adenosine diphosphate. Double beam spectrophotometric trace of DPNH oxidation (an upward deflection corresponding to oxidation) and platinum microelectrode recording. Time proceeds from left to right, oxygen utilization is indicated by an upward deflection of a platinum electrode trace. The addition of the reactants is indicated on the trace as are the rates of oxygen utilization (μM/sec) and the amounts of oxygen used up following ADP addition and following calcium addition. The reaction media is approximately 0.1 M NaCl with .015 M phosphate, pH 7.2, magnesium and fluoride being omitted, temperature 26°, approximately 12 μM DPNH. (Expt. 466C).

A number of observations were made. Firstly, respiration was stimulated - not inhibited - by Ca^{++} and beyond the level obtained with ADP and phosphate by ratio of 1.7:1. Secondly, the amount of oxygen utilization was limited, i.e., the effect of calcium was reversible, presumably dependent upon its expenditure in the reaction involving its binding to the mitochondria as described by Slater and Cleland (3). Thirdly, the ADP/O_2 value for the first addition is 6.8. The $calcium/O_2$ value for the second addition is 15. In other words, 2.2 calcium are required to impose an energy load upon the mitochondria equivalent to 1 ADP. Fourthly, independent experiments showed that the reaction of calcium with the mitochondria was not dependent upon ATP formed from ADP phosphorylation, as might be inferred from this figure; similar results were obtained without the prior addition of ADP. (See Fig. 4 of ref. 4). Lastly, these experimental results are consistent with those of previous workers who used high concentrations of Ca^{++} and observed respiration inhibition and irreversible swelling.

DeLuca and Engstrom (8) and Vasington and Murphy (9) have verified the earlier experimental results, but have concentrated their attention upon the effects of higher concentrations of Ca^{++} and upon the maximal Ca^{++} uptake of rat kidney mitochondria. The former workers noted the significant lack of oligomycin inhibition of the reaction; leading to the current view that internal high energy intermediates as well as ATP itself may act as energy sources for the ion accumulation (10-12). Current work is reported in the previous two papers (10, 11) and at the Federation Meeting (12-14).

Much of the current work is directed towards showing the unusual extent to which calcium is accumulated by the mitochondria. These results were however foreshadowed by the study of Slater and Cleland (3) who showed that versene-treated acid-washed sarcosomes took up 10^{-4} molar calcium; in fact, they stated that in the usual preparation "isolated sarcosomes contain all the calcium in the heart muscle." Thus the idea of a tight binding of calcium with the mitochondrial membranes was well established by their pioneer work. The fact that binding of manganese to the mitochondrial membranes can be elegantly studied by the measurements of proton relaxation times, according to the spin echo method (11), gives testimony to the validity of the earlier conclusions.

Observations of the large increase in the hydrogen ion concentration concomitant with calcium binding can be conveniently followed with sensitive recording by means of a glass electrode (15, 16) and this observation has also been made by Brierley et al (14). The role of phosphate in the calcium uptake has received particular attention in Lehninger and Green's laboratory (13, 14) and the formation of calcium phosphate precipitates appears clearly established by electron microscopy.

Attention is now focused on the key role of calcium in the electrogenic properties of axons (17) and in muscular contraction as well (8). Thus any mechanism for the binding and liberation of calcium from intracellular structures is of considerable interest in functional tissues.

ENERGY-LINKED FUNCTIONS OF MITOCHONDRIA

Chappell's arrival here for an interval of research prior to the Federation meeting has stimulated me to carry out with him some revisitations of the calcium-activated respiration experiments. I am happy to present some more recent experiments in which calcium has stimulated the respiration as indeed Chappell has stimulated me.

Responses of respiration, cytochrome b and reduced pyridine nucleotide to low concentrations of calcium.

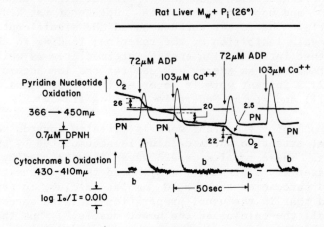

Fig. 2. Comparison of ADP and calcium-stimulated respiration in rat liver mitochondria. Oxygen utilization measured by the vibrating platinum microelectrode, cytochrome b measured by the double beam spectrophotometer, and pyridine nucleotide oxidation measured fluorometrically. 2 mg protein rat liver mitochondria suspended in 80 mM KCl, 20 mM Tris-Cl, 5 mM P_i, 5 mM succinate and 5 mM glutamate (final concentrations). Total volume 3.0 ml., temperature 26°, pH 7.5. (Expt. 782 A-III).

Fig. 2 illustrates the response of cytochrome b and reduced pyridine nucleotide to calcium activation of respiration in liver mitochondria. In this study low concentrations of calcium are employed so that the repeated response to this substance can be recorded. Cytochrome b oxidation is indicated by an upward deflection of the trace. Pyridine nucleotide is recorded by fluorometric excitation at 366 mμ with measurement of emission at 450 mμ. An upward deflection of the trace corresponds to an oxidation of reduced pyridine nucleotide. Oxygen is registered polarographically by means of the vibrating platinum microelectrode; in contrast to Fig. 1 a decrease of oxygen concentration is registered with a downward deflection (see Ref. 20 for a more complete description of the apparatus). The advantages of the vibrating platinum microelectrode for the rapid changes of oxygen concentration occurring on addition of such low calcium concentrations of mitochondria need scarcely be emphasized; electrodes covered with teflon or polyethylene membranes may well respond too slowly to properly record the kinetics of oxygen utilization of the type shown in Fig. 2. In agreement with Davies and Brink (19), we have found a thin film of collodium to permit a rapid response and yet to give a high stability (19). The rat liver mitochondria are suspended in a phosphate-containing reaction medium and supplemented with succinate and glutamate; initially the mitochondria are slowly respiring in State 4 (left-hand portion of the figure). Addition of 72 μM ADP accelerates respiration to 4.2 μM oxygen/sec and causes nearly synchronous cycles of oxidation and reduction of cytochrome b and pyridine nucleotide. A repetition of additions of ADP and calcium cause repetitions of the cyclic responses. Under these conditions it may be seen that the phosphorylation of ADP and the uptake of calcium is a stable phenomenon and may be repeated at will. The oxygen uptake in the interval of stimulated respiration is 26 and 20 μM oxygen for additions of 72 μM ADP and 103 μM calcium, corresponding respectively to ADP/O_2 values of 2.8 and $calcium/O_2$ values of 5.2 The ratio of these two quantities indicates that 1.8 calciums can be bound/ADP phosphorylated to ATP. This value is in reasonable accord with the value 2.2, obtained in the earlier experiment (Fig. 1).

Affinity for calcium. One of the main reasons for the great interest in calcium binding by mitochondria is that it may provide a model for active transport under physiological conditions. In view of the very tight binding of calcium by components of the endoplasmic reticulum (18) - less than 0.01 μM calcium may be free - it is of considerable importance to determine how much calcium is required to obtain responses of respiration rate and carrier oxidation-reduction. Since usual methods for cation analysis (including "EPR") are of inadequate sensitivity to determine affinity constants with precision, we have utilized the sensitive method illustrated by Fig. 2 for this purpose. An example of the titration of rat liver mitochondria with calcium is provided by Fig. 3. Glutamate and succinate are used as substrate. In order to ensure that the reaction is drive by the internal high energy intermediate $(X \sim I)$, oligomycin is added (11 μg/ml). In the presence of oligomycin maximum values of the

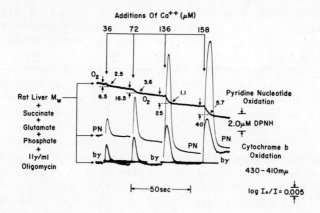

Fig. 3. Titration of respiratory and electron carrier responses to additions of calcium. 2.3 mg/ml rat liver mitochondria suspended in Tris KCl medium at 26° and supplemented with 4 mM succinate and glutamate, 3.6 mM phosphate, and 11 μg/ml oligomycin. (Expt. 784-5).

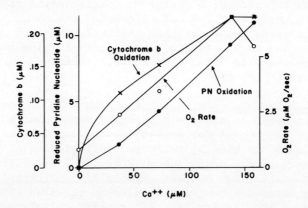

Fig. 4. Graph of the experimental data of Fig. 3; increment of oxygen uptake and increment of pyridine nucleotide and cytochrome b oxidation plotted as a function of added calcium concentration. Other conditions as in Fig. 3. (Expt. 784-5)

respiratory control ratios are also obtained. The first addition of 36 μM calcium causes cyclic oxidation-reduction responses of cytochrome b and pyridine nucleotide together with a measurable amount of rapid oxygen uptake. Addition of

72 µM calcium gives a more readily measurable oxygen utilization and larger and longer cyclic responses of the carriers. Respiration is further accelerated at 136 µM but at 158 µM calcium the respiration rate has apparently reached a saturation level. However, the response of pyridine nucleotide is still increasing with calcium concentration. In Fig. 4 the magnitudes of the responses of the cytochrome b and pyridine nucleotide are plotted together with the magnitude of the respiratory rate, as a function of calcium concentration. At 136 µM calcium both the cytochrome b response and the respiratory rate have reached maximal values; in fact, at 158 µM calcium the respiratory rate is slightly inhibited. Both traces are clearly non-hyperbolic and thus the value for 50 µM or half-maximal effect is undoubtedly a minimum value. Nevertheless, the graphs indicate that the State 4 respiration rate is doubled by 25 µM calcium, a rather high affinity of the mitochondria. These values for calcium may be contrasted with the low affinity for manganese where 750 µM gives half-maximal respiratory stimulation (11). The response of reduced pyridine nucleotide still seems to be increasing at calcium concentrations which cause saturation in the response of cytochrome b and respiratory activity. Inasmuch as we have employed rat liver mitochondria in this experiment, both TPNH and DPNH are being oxidized. TPNH oxidation, which must pass through transhydrogenase activity (21, 22) may well interfere with measurements of cycles of DPNH oxidation; thus we regard observation of cytochrome b as more accurate.

With pigeon heart mitochondria it is desirable to compare the responses of the system in the presence and absence of cyclic responses of pyridine nucleotide; the experiment is carried out in the presence and absence of Amytal. The amplitude of the response of cytochrome b is plotted as a function of calcium concentration in Fig. 5. Responses are recorded in the range of 14 to 158 µM calcium. Half-maximal effect is obtained at 45 µM calcium in the absence of Amytal and 40 µM calcium in the presence of Amytal. Thus the calcium concentration for half-maximal response is nearly independent of the presence or absence of Amytal. The values 40-45 µM calcium correspond to a "Michaelis constant" for the reaction of calcium with the mitochondria and do not indicate the value of an "equilibrium constant." Judging from the "cycles" of cytochrome b with 14 µM calcium, it is probable that the amount of calcium that could be in equilibrium with mitochondria is very small compared to the 40 µM "Michaelis constant", a few µM Ca^{++}.

Role of high energy DPNH in calcium accumulation. Since there has been much discussion of the possibility of the formation of high energy forms of DPNH by reversed electron transfer, it is of interest to determine whether concentrations of calcium of the same order as the intramitochondrial DPNH concentration can react more rapidly with mitochondria than larger concentrations of calcium. This experiment resembles the ATP-jump experiment (23) except that calcium is a more rapidly reacting energy acceptor than is ADP (24). We find it convenient to have computed the rate of calcium uptake from the formula (25)

$$k_3 = Ca^{++}/P_{max} t_{1/2} \text{off} \qquad (2)$$

This equation originally was developed for calculating turnover rates of substrates in a simple enzymatic system; it can satisfactorily be applied to the mitochondria as evidenced from the response to known amounts of ADP (26). In effect, this equation correlates the amount of calcium added, the extent of cytochrome b oxidation, and the duration of the stimulated respiration to give a quantity proportional to the turnover rate of calcium.

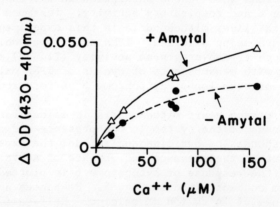

Fig. 5. Response of cytochrome b of pigeon heart mitochondria to additions of calcium in the presence and absence of Amytal. 1.65 mg/ml mitochondrial protein suspended in mannitol-sucrose-tris medium, pH 7.4, temperature 26°. Mitochondria supplemented with 4 mM glutamate and succinate, 3.6 mM phosphate, and 11 μg oligomycin/ml; in the plus Amytal curve the concentration is 1.4 mM Amytal, 26°, pH 7.4. (Expt. 785-II-6).

Cycles of this type were obtained in the studies of Fig. 5 and the rate of reaction in calcium binding studied in the presence or absence of Amytal, and is evaluated from the response of cytochrome b. The results are plotted in Fig. 6. In the top trace the lowest concentration of calcium (14 μM)

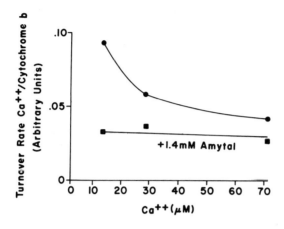

Fig. 6. The effect of Amytal upon the rate of reaction of calcium with the mitochondria. k_3 (see Eq. 2) computed from cycles of oxidation and reduction of cytochrome b in response to additions of calcium. Experimental conditions identical to those of Fig. 5. (Expt. 785).

gives a rate that is nearly twice that obtained at higher concentrations of calcium. It is possible that the ratio of the rates is even greater at lower concentrations of calcium. In the presence of Amytal there is no evidence of an initially higher rate of calcium uptake at low calcium concentrations. Thus the kinetic data suggests that two entirely different rate processes occur - on the one hand, in the presence of DPNH oxidation and, on the other hand, in the absence of DPNH oxidation - since spectroscopic observations show a large response of reduced pyridine nucleotide in the absence of Amytal and a negligible response in the presence of Amytal. We believe our experimental results point to the possibility of DPNH as an energy donor in the rapid uptake of small concentrations of calcium.

The amount of calcium taken up very rapidly in the absence of Amytal is 20-30 μM. This may be compared with the amount of DPNH present in the mitochondria (about 16 μM). It would seem, therefore, that the DPNH compound can act as an energy donor binding about 2 equivalents of calcium. The binding of larger amounts of calcium presumably requires the turnover of the respiratory carriers and proceeds at a slower rate.

The effect of phosphate. The preceding papers have called particular attention to the role of phosphate in calcium and manganese accumulation and brought about a revival of interest in Slater and Cleland's observation that phosphate was not necessary for calcium binding to sarcosomes (3). Pigeon heart mitochondria are especially suitable for this purpose in view of their low content of endogenous phosphate (it has been estimated that the phosphate content may be as low as 1/cytochrome c in pigeon heart mitochondrial preparations (20)).

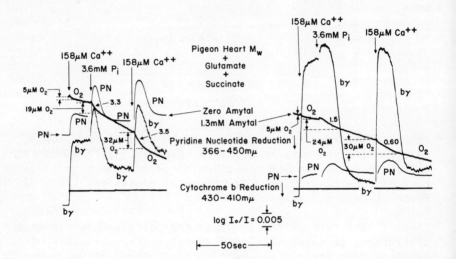

Fig. 7. Illustration of the response to calcium in the absence of phosphate in Amytal-free and Amytal-pre-treated pigeon heart mitochondria. 1.5 mg protein/ml pigeon heart mitochondria suspended in MST medium and supplemented with 4 mM glutamate and 4 mM succinate. Traces on the left are recorded in the absence of Amytal and on the right in the presence of 1.3 mM Amytal. Time proceeds from left to right on both traces and cytochrome and pyridine nucleotide reduction are indicated as a downward deflection. The platinum microelectrode traces include the increments of oxygen concentration and the rate (μmolar/sec) 26°, pH 7.4. (Expt. 778-III,5).

The pigeon heart mitochondria, supplemented with succinate and glutamate, respond to the addition of 158 µM calcium with a jump upward to a more oxidized steady state. Two phenomena distinguish this response from that obtained in the presence of phosphate: first, the increment of oxygen uptake is scarcely measurable ($\sim$ 5 µM); second, the oxidation of the steady state levels is stable, i.e., no cyclic response of the respiratory carriers is observed as in the previous records where phosphate was present. We may interpret this record to mean that reaction with calcium does not go to completion under these circumstances and free calcium is still present- either in the reaction medium or in equilibrium with a binding site of low affinity in the mitochondria. This observation receives direct support from the response to the addition of 3.6mM phosphate which immediately elicits rapid respiration (3.3 μM/sec) until 19 μM oxygen is expended. Furthermore, on addition of phosphate the respiratory carriers jump to an even more oxidized steady state, then return to a reduced steady state as the respiratory activity ceased. The response to the second addition of 158 μM calcium is typical of phosphate-supplemented mitochondria; 32 μM oxygen is taken up at the rate of 3.5 μM/sec and the response of the oxidation-reduction cycles to the respiratory carriers is closely synchronized with the respiratory activity.

The effect of calcium upon the steady states of cytochrome b and upon the respiratory activity in pigeon heart mitochondria to which 1.3 mM Amytal has been added is indicated in the right-hand portion of the figure. Addition of 158 μM calcium causes an almost complete oxidation of cytochrome b (compare the upward deflection of the traces in the left-hand portion of Figure 7). Again, a small amount of oxygen is expended (about 5 μM). Addition of phosphate causes only a small increment in oxidation of cytochrome b and acceleration of respiration to 1.5 μM/sec, as well as the utilization of 24 μM oxygen. A small response of pyridine nucleotide is observed, no doubt due to incomplete inhibition of reduced pyridine nucleotide oxidation by the Amytal concentration employed. The second addition of calcium gives a response which is typical, i.e., 30 μM oxygen is used at a rate of 0.6 μM/sec. The decrease of oxygen rate is probably due to the higher sensitivity of the Amytal-treated mitochondria to calcium inhibition of respiration. Again the requirement of phosphate for the completion of the reaction with calcium is indicated.

Discussion.

Is active transport involved? Superficially, the interaction of divalent cations with mitochondria bears all the "earmarks" of an active transport process; a reaction with ATP itself or with high energy intermediates of the mitochondria is supported by various data. In fact, data of this paper indicate that intermediates with which calcium may interact may be $X \sim I$, $DPNH \sim I$, or $DPNH \sim P$. The accumulation of ions against the thermodynamic gradient across a semi-permeable membrane should also lead to the development of a membrane potential but unfortunately this will remain an unknown quantity until membrane potentials in mitochondria can be measured directly. As discussed below, an ATPase activity (or $X \sim I$ase) may accompany the binding of calcium to the mitochondrial structure in which energy of ATP is not required to form the final reaction product. However, the most likely role of energy-linked intermediates in calcium binding would be the accumulation of sufficiently high concentrations of calcium within the mitochondrial membrane to cause precipitation of calcium phosphate.

Reaction Sequence. These experiments identify three steps in the reaction of mitochondria with calcium. The first reaction is a very rapid one (4) and is phosphate independent; the respiratory carriers reach highly oxidized steady states, nearly completely oxidized in the Amytal-treated mitochondria (inhibited State 3). This phase corresponds to the first phase of manganese binding (11) followed by EPR and by spin echo methods. It should be noted that this step, in agreement with Chappell, is not an energy-utilizing process, i.e., the mitochondria do not respire steadily (except for an initial burst) in the presence of calcium and in the absence of phosphate. This step is oligomycin insensitive (11). The second step occurs upon addition of phosphate; a rapid energy-linked process occurs which stimulates respiration and causes highly oxidized states of the respiratory carriers (State 3). In the third step Ca^{++} is bound in a form that no longer stimulates respiration and the respiratory carriers return to the resting state (State 4). In this state a calcium is tightly bound, corresponding to a low enhancement value of the spin echo technique.

Reaction mechanism. A reaction mechanism, which incorporates many of the known aspects of calcium and manganese binding in transport (11) is indicated in Fig. 8. The first step indicates binding of calcium to cristae of the mitochondria. It is assumed that entry of calcium into the matrices of the mitochondria can occur by passive diffusion. This fast step is not energy dependent and does not cause changes in the carrier oxidation states. The spin echo method indicates a decrease of proton relaxation time when manganese is used. This may, for example, be a result of a large magnetic effect of manganese on the protons of water and analogous to

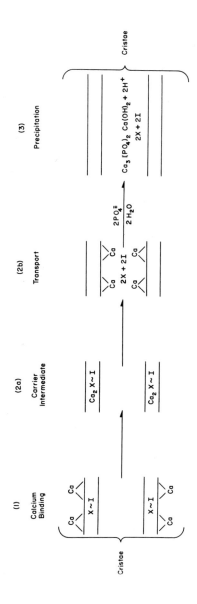

Fig. 8. Four steps in the reaction of calcium with the cristae of the mitochondria. (MD 150).

the binding of manganese to a protein of the outer surface of the cristae. This suggestion also is consistent with the views of Klingenberg, who points out that a sufficient area for the initial phase of divalent cation binding can only be found on the more extended surfaces of the cristae as compared with the area of the outer membrane of the mitochondria (24). The second step is energy-linked but does not involve phosphate. The spin echo data indicate that proton relaxation time has considerably increased (decreased enhancement) and the spectroscopic data indicate that increased oxidation of the respiratory carriers occurred but without sustained respiration (see above). In this second stage the diagram indicates that divalent cations combine at the same site as ATP, guanidine, magnesium ($X \sim I$) and two calcium ions are indicated to combine with $X \sim I$ to satisfy the observed stoichiometry. Protons may be released at this stage also (27). Since the response to calcium is sustained, i.e., calcium does not appear to be reduced to a concentration where the carriers return to their reduced steady state, it must be presumed some calcium remains combined with the "carrier" molecule (2a) - presumably because the dissociation constant for the calcium receptor at this stage is rather high.

It is only in the third step that calcium is bound so tightly that the proton relaxation time increases to its maximal value (minimal enhancement) and the respiratory carriers return to the reduced steady state - that obtained before calcium addition. This step requires phosphate and presumably leads to the precipitation of insoluble calcium phosphate. It is accompanied by a swelling of the mitochondrial membranes (a diminution of fluorescence and an increase of light-scattering).

This paper identifies one of the efficient energy donors for the third stage as reduced pyridine nucleotide, possibly as a high energy intermediate $DPNH \sim I$ or $DPNH \sim P$. It will be interesting to determine whether or not this intermediate can donate energy directly in the combining reaction or in the calcium binding reaction.

In the fourth step, excess Ca^{++} can damage the mitochondrial membranes (see Fig. 1).

ENERGY-LINKED FUNCTIONS OF MITOCHONDRIA

SUMMARY

The transient activation of respiration of tightly coupled mitochondria by addition of low concentrations of calcium (less than 300 µM) is a phenomenon that clearly identifies the energy requirements of the reaction of calcium and mitochondria. The energy requirements for calcium binding published previously (4) agree closely with those determined by more recent studies: approximately two calciums are equivalent to 1 ADP. The affinity for calcium is high (under 100 µM) in the case of rat liver mitochondria and 45 µM in the case of pigeon heart mitochondria. Only extremely low concentrations of calcium could be in equilibrium with mitochondria.

An investigation of the rate of uptake of very small concentrations of calcium (10-20 µM) indicates that small amounts are taken up over 2 times as rapidly as larger amounts (100 µM). Furthermore, this extremely rapid uptake of small concentrations of calcium is abolished by the presence of Amytal. For this reason, the high energy intermediate of the first phosphorylation site (DPNH $\sim$ I or DPNH $\sim$ P) seems to be a particularly effective energy donor in calcium accumulation in mitochondria.

A comparison of the response of respiratory carriers in the absence and presence of phosphate reveals three steps in the reaction in addition to the energy independent step observed by the spin echo technique: (2a) the reaction of calcium with the mitochondria which leads to a highly oxidized steady state in the carriers but low rates of respiration. Presumably calcium eneters the mitochondria and is bound to a high energy intermediate; (2b) the energy-requiring step in calcium uptake in which respiration is rapid, the carriers are in State 3; the light-scattering is increased. This stage occurs only in the presence of phosphate in pigeon heart mitochondria; (3) the step in which the added calcium has been bound by the mitochondria (presumably in the form of calcium phosphate); the respiration is low (State 4); the carriers are highly reduced; (4) in the presence of excess calcium the light-scattering rapidly increases; reduced pyridine nucleotide (probably cytochrome b) is highly oxidized, and respiration is highly inhibited.

ACKNOWLEDGEMENT.

The author would like to express his appreciation and gratitude for the early contributions of Dr. G.R. Williams to this work in 1955 and more recently to Dr. Brian Chappell for his visit here and his work on manganese accumulation by mitochondria, which stimulated further interest in calcium accumulation by mitochondria. Also many thanks are due Reiko Oshino for mitochondrial preparations.

This research was supported in part by a grant from the U.S. Public Health Service.

REFERENCES

1. Siekevitz, P. and Potter, V.R., J.B.C., 201, 1 (1953).
2. Potter, V.R., Siekevitz, P. and Simonsen, H.G., J.B.C., 215, 893 (1953).
3. Slater, E.C. and Cleland, K.W., Biochem. J., 55, 566 (1953).
4. Chance, B., Proc. III Intern. Congr. Biochem., Brussels, 1955, Academic Press, N.Y., 1956, p. 300.
5. Chance, B., and Williams, G.R., J.B. C., 217, 383 (1955).
6. Chance, B., and Williams, G.R., J. B. C., 217, 409 (1955).
7. Chance, B., Science, 120, 767 (1954).
8. DeLuca, H.F. and Engstrom, G.W., Proc. Nat. Acad. Sci., 47, 1744 (1961).
9. Vasington, S.D. and Murphy, J.V., J.B.C., 237, 2670 (1962).
10. Brierley, G.P., this volume, p. 237.
11. Chappell, J.B., this volume, p. 224.
12. Chappell, J.B. and Greville, G.D., Fed. Proc., 22, 526 (1963).
13. Lehninger, A.L., Rossi, C.S. and Greenawalt, J., Fed. Proc., 22, 526 (1963).
14. Brierley, G.P., Murer, E., Bachmann, E. and Green, D.E., Fed. Proc., 22, 526 (1963).
15. Chance, B. and Hollunger, G., Abs. Am. Chem. Soc., N.Y. Ann. Meet., Sept. 1957, p. 43C.
16. Chance, B. and Hollunger, G., J.B.C., 238, 439 (1963).

17. Frankenhaeuser, B. and Hodgkin, A.L., J. Physiol., 137, 217 (1957).
18. Weber, A., Hertz, R. and Reiss, I., J. Gen. Physiol., 46, 679 (1963).
19. Davies, P.W. and Brink, F., Rev. Sci. Instr., 13, 524 (1942).
20. Chance, B. and Hagihara, B., Proc. Vth Internatl. Cong. Biochem., Moscow, 1961, Pergamon Press (in press).
21. Danielson, L., this volume, p. 157.
22. Hommes, F.A., this volume, p. 39.
23. Schachinger, L., Eisenhardt, R. and Chance, B., Biochem. Z., 333, 182 (1960).
24. Klingenberg, M., this volume, p. 121.
25. Chance, B., J.B.C., 151, 553 (1943).
26. Chance, B. and Williams, G.R., J.B.C., 221, 477 (1956).
27. Chance, B., Pressman, B.C. and Saris, N-E., unpublished observations.

DISCUSSION

Packer: Some years ago, I observed the effects of calcium on light-scattering properties and respiration of mitochondria. I noticed a decrease in the electrode current when calcium was added to the solution in the absence of mitochondria. Have you observed this?

Chance: We are using collodion-covered electrodes.

Packer: If the electrode is collodion-covered, then I guess you are all right. During these short bursts of oxidation and reduction of the respiratory carriers in the titration with calcium, does the structure of the mitochondria change?

Chance: The light-scattering changes synchronously with the additions of low concentrations of calcium, just as it did in our original experiments with ADP.

Packer: May I propose a possible alternate explanation for these phenomena? Suppose the calcium activates ATPase, then the increase in the ADP level would cause an oxidation of the carriers. But then the calcium is transported and becomes insoluble - it can no longer activate the ATPase, and the respiratory carriers return to their original level.

Chance: But the reaction is not ATP-dependent under our conditions.

Chappell: Also, the reaction is oligomycin-insensitive.

Packer: But heart mitochondria have an active calcium- and magnesium-activated ATPase.

Chappell: Yes, but the calcium-activated ATPase is inhibited by oligomycin. This is calcium transport, and the effect upon the respiratory carriers is oligomycin-insensitive.

Slater: It has long been known that, while calcium destroys DPN in homogenates, it has no effect on isolated DPN-destroying enzymes. No one has really found out why calcium is so effective in destroying DPN in homogenates and in the mitochondrial preparations.

Packer: Is the DPN so completely destroyed by hydrolyzing enzymes? This first requires the release of DPN.

ENERGY-LINKED FUNCTIONS OF MITOCHONDRIA

<u>Slater</u>: Yes, we (1) suggested that too, but perhaps there is a more specific mechanism for breaking down DPN.

<u>Klingenberg</u>: I should like to show two figures on the incorporation of calcium, magnesium, and phosphate in locust mitochondria, from our studies several years ago. These mitochondria have the glycerol phosphate-linked, energy-dependent DPN reduction.

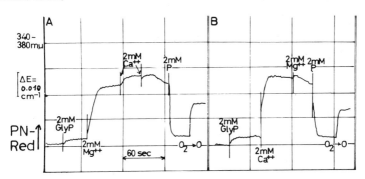

Fig. 1. The influence of phosphate on ion transport linked redox state of the reapiratory chain. Recording of pyridine nucleotide absorption in a suspension of locust mitochondria. The energy linked reduction of DPN by glycerol phosphate is initiated on addition of Mg^{++} (A) or Ca^{++} (B) since glycerol phosphate dehydrogenation is activated by these ions. Oxidation is effected through the further addition of phosphate only. On anaerobiosis the pyridine nucleotides remain oxidized. The increase of absorption is due to cytochrome c interference. (Klingenberg)

Figure 1 shows the reduction of DPN which is initiated by magnesium, since glycerol phosphate oxidase requires magnesium. Subsequent additions of calcium are without an immediate effect. Only phosphate then causes extensive oxidation. In a second, somewhat paradoxical, experiment the energy-dependent reduction of DPN is initiated by calcium, which may eventually release internally-bound magnesium. The oxidation of DPNH which reflects the "uncoupling" action of calcium only occurs again together with phosphate.

Figure 2 shows the same effects for flavin-non-hemin iron absorption. It is reduced (here, a decrease of absorption) only after addition of magnesium and phosphate, and oxidized again with calcium. The reduction on anaerobiosis is due to cytochrome a_3 interference. Further, the reduction may be inhibited by addition of calcium and again oxidized by phosphate addition.

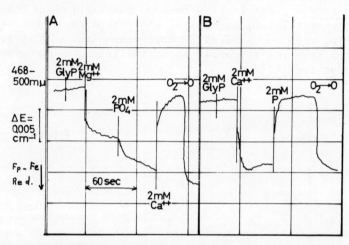

Fig. 2. Recording of flavin-Fe-absorption in a suspension of locust mitochondria. The reduction of flavin-Fe can be initiated either by Mg^{++} or Ca^{++} and is abolished only after Ca^{++} and phosphate are added. In anaerobiosis flavin-Fe remains oxidized. The decrease of absorption is due to cytochrome a_3 interference. (Klingenberg)

These experiments demonstrate the energy supply, since the reduction of DPN and flavin is here shown to be energy-dependent (2). Calcium, together with phosphate, can "uncouple", i.e., "deflect", energy - probably for ion transport. Magnesium is not efficient in this way.

In liver mitochondria the addition of calcium also gives extensive oxidation, again only in the presence of phosphate as demonstrated in Fig. 3. With ketoglutarate, this extensive oxidation is accompanied first by a burst, and then by an inhibition of respiration. With succinate, the respiration remains activated (not shown). This is interpreted as the removal of DPN from the dehydrogenases. Thus there may be first a respiration-supported transport which is then abolished due to an (irreversible) inhibition of the energy transfer.

With a new mixing device we have studied the velocity of DPNH oxidation through addition of an excessive concentration of calcium in the presence of phosphate, and compared this to the rate of oxidation with ADP, in order to find some hint

ENERGY-LINKED FUNCTIONS OF MITOCHONDRIA

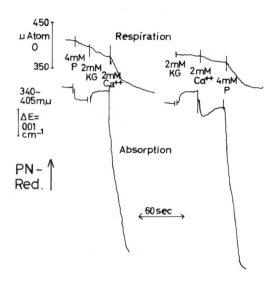

Fig. 3. The interdependence of calcium and phosphate effects on the respiratory chain. Simultaneous recording of pyridine nucleotide absorption and oxygen concentration in a suspension of liver mitochondria. The addition of phosphate or calcium alone has only minor effects, whereas on the combined addition a rapid oxidation of the pyridine nucleotides and a transient stimulation of the oxygen uptake is recorded. (Klingenberg)

about the point of action of calcium compared with that of ADP. This means that the turnover in the respiratory chain when stimulated with calcium is five times higher than on the transsition to oxidative phosphorylation. This would be evidence that calcium is acting at a point prior to ADP in the energy transfer sequence.

We believe that calcium is going from the outer to the inner space of the cristae, accompanied by phosphate, which acts by anion exchange with the internal anions, otherwise found in the inner membrane. Thus pyridine and adenine nucleotides are released and then, for example, DPN can be easily removed and even become accessible to DPNase. This, of course, is at the time of the exchange reaction with the proton transport of the mitochondria.

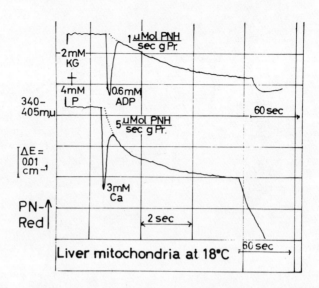

Fig. 4. The release of reduced pyridine nucleotides to the respiratory chain by ion transport. Comparison of the rates of pyridine nucleotide oxidation on transition from the controlled state to the ADP activated or Ca^{++}-activated state of mitochondria. Recording of pyridine nucleotide absorption in liver mitochondria. Addition of ADP and Ca^{++} with a rapid mixing device. KG = ketoglutarate, PNH = reduced pyridine nucleotides. (Klingenberg)

Chappell: In view of the large quantities of manganese transported into liver mitochondria in the absence of phosphate, I do not believe that your theory, if I have understood it correctly, is defensible. The manganese is far too high; 300 μmoles manganese and only 20 μmoles or so phosphate.

Klingenberg: It is high; but there are enough phospholipids in the mitochondria which are probably in the cristae - you gave the number in your paper. From the high rate of DPNH oxidation after adding calcium plus phosphate, it may be concluded that the DPNH is in a compartment where it primarily is inaccessible to oxidation, since it is much more rapidly released from this state by calcium than by ADP. This shows that the release of DPNH is more a transport process than merely a removing of the energy-rich intermediate forms.

REFERENCES

1. Slater, E.C. and Cleland, K.W., Biochem. J., <u>55</u>, 566 (1953).
2. Klingenberg, M. and Bücher, Th., Bio. Z., 334 (1961).

ENVOI

Chance: In conclusion, I should like to thank the people who have come here to make this fine colloquium, and also those from the Johnson Foundation who have helped see that the meeting here has run smoothly. It has been a great pleasure to have you all here.

Slater: On behalf of all the visitors, I should like to thank you very much for organizing this colloquium. We are used to hearing about your rapid reactions, but this is certainly an extremely rapidly organized colloquium, and one with an immense yield. I think everyone who has ever done an experiment on the reversal of the respiratory chain is here. We are very grateful indeed to have heard the very latest in this field, even those things which don't even get published in the BBRC, but are exchanged by telephone.

INDEX OF PARTICIPANTS

Brierley, G.P.; 237-245, 83, 246, 249, 250.

Chance, B.; 253-269, 1-2, 17, 18, 19, 20, 21, 22, 25, 37, 72, 73, 82, 93, 115, 118, 177, 178, 180, 200, 203, 249, 250, 270, 275.

Chappell, J.B.; 219-231, 83, 200, 201, 202, 233, 234, 235, 246, 250, 270, 274.

Cohn, M.; 179, 180, 233.

Conover, T.C.; 24, 37, 155, 178.

Danielson, L.; 157-175, 176, 178, 179.

Davies, H.; 72.

Estabrook, R.W.; 143-152, 22, 23, 25, 36, 37, 38, 73, 82, 84, 114, 115, 118, 119, 153, 154, 155, 156, 177, 178, 179, 180.

Green, D.E.; 120, 141, 202, 203, 232, 233.

Griffiths, D.; 49, 93, 115, 116, 117, 118, 119, 120, 155.

Hess, B.; 17, 72, 93, 118.

Hommes, F.A.; 39-48, 17, 23, 37, 49, 234.

Jöbsis, F.; 249.

King, T.E.; 36, 82, 83, 84.

Klingenberg, M.; 121-139, 18, 22, 23, 92, 93, 114, 140, 141, 154, 155, 177, 180, 232, 234, 235, 246, 247, 250, 271, 272, 273, 274.

Löw, H.; 5-16, 17, 18, 22, 23, 24, 36, 37, 114, 235.

Mildvan, A.; 22, 118, 156.

Packer, L.; 51-71, 23, 24, 25, 72, 73, 92, 93, 234, 270.

Penefsky, H.; 87-91, 92, 93.

Pressman, B.; 181-199, 119, 120, 140, 141, 200, 201, 202, 203, 246, 248, 249.

Pullman, M.; 24, 36, 84.

Racker, E.; 75-81, 17, 22, 23, 24, 36, 82, 83, 84, 176, 177, 178, 180, 200, 233, 234.

Roy, B.H.; 93.

Sanadi, R.; 26-35, 23, 36, 37, 38, 83, 114, 153, 155.

Slater, E.C. 97-113, 21, 22, 23, 24, 37, 72, 73, 93, 114, 115, 117, 118, 140, 176, 179, 180, 270, 271, 275.

Webster, G.; 19, 83, 84.

INDEX TO SUBJECTS

active transport, 234; H^+ release, 197

Amytal, 104, 106, 107, 221; effect on ketoglutarate, 127; effect on transhydrogenase, 160

antimycin A, 214, 221; effect energy-linked reactions, 98, 106, 107, 160; effect on H^+ release, 197; effect on Ubiquinone, 92; inhibition: of ATPase, 19-22; of DPNH oxidation by quinone, 32, 36, 92; of ion accumulation, Mn^{++}, 223, 225; Mg^{++}, 237, 240; of pyridine nucleotide reduction, succinate linked, 6, 19; dye-linked, 6, 7; quinone-linked, 27, 28; insensitivity to, 28, 37

arsenate, 106

atractylate (atractyloside), 17-18; 164

bovine serum albumin: effect on transhydrogenase, 155

CPP: effect of guanidine inhibition, 191

chloroplasts: energy-linked reactions, 51; shrinking-swelling, 62

coupling factors: ATPase activity, 31, 36; cold lability, 36; in DPN reduction, 29-30; in oxidative phosphorylation, 30; topography of, 75

coupling factor F_1 (ATPase): 13, 22, 23, 177; F_4, 177

creatine kinase, 141

dicumarol: uncoupling effect, 183; with octylguanidine, 183; loci of action, 184

dinitrophenol: effect on energy-linked H^+ transfer, 124, 125, 166; on H^+ migration, 196, 203; on guanidine inhibition, 182; on DPNH $\sim$ P, 117; on succinate-linked pyridine nucleotide reduction, 106, 166, 172; on transhydrogenase, 145, 147, 155; general effects, 181

directional enzymes, 137

DPN $\sim$ I (NAD $\sim$ I): 106, 110, 115

DPNH $\sim$ P (NADH $\sim$ P): 117; structure, 118; relation to DPNH-X, 119; intermediate in oxidative phosphorylation, 172

DPNH-X: 116, 119

EDTA, effect on transhydrogenase, 161, 162

electron flow, non-cyclic, 207, 208

electron transfer, cyclic, 208; reverse, 7, 154, 207

electron spin resonance (EPR), 224

flavoprotein, DPNH-CoQ_o reductase, 33-35, 37, 38

fumarate, reduction by NADH, 111, 114; effect of uncoupler, 211; of oligomycin, 213; light-dependent reduction, 209, 210

glutamate formation, energy-linked: from malate, 107, 123, 131; from isocitrate, 108; effect of phosphate on, 132, 140

glyceraldehyde-3-phosphate: reaction, 97; form of NAD, 110, 114

glycerol phosphate, 97

Gramicidin, 191

guanidines: 182; effect of ADP, 201; effect on cytochrome b reduction, 187; hexyl, 104; octyl, 182; others, 185; loci of action, 184-186; relation to oligomycin, 192; to uncouplers, 200

histidine phosphate, in ETPH, 19

hydrogen ion transport, 135; during ion uptake, 202

hydrogen pathways; dynamic organization, 121, 135; energy control, 122, 130

hydrogen, photoproduction of, 214; effects of uncoupling agents, 215

intermediates, energy-rich, 211, 213, 214, 216

ion accumulation, 219; Mn^{++}, 219, 220; Ca^{++}, 253, 271

magnesium: activates NTPase, 171; α-ketoglutarate dismutation, inhibition of, 127; binding constant with ATP and ITP, 18; release of H^+, 202; inhibition of DPN reduction, 25; transhydrogenase, effect on, 153, 161, 178; effect of valinomycin on active transport of, 195

menadione (vitamin K_3), DPN reduction by, 26

mitochondria, swelling, 59

nucleotides: CTP, 13, 14; GTP, 13, 14; ITP, 11-13, 14, UTP, 13, 14

NTPase: effect of on oligomycin, 163, 165; relation to energy transfer, 171, 172

oligomycin, 213, 214, 216, 221, 222, 223; comparison with guanidines, 183, 192; effect on energy-linked reactions, 100, 103, 124, 154, 155, 160, 161, 163, 171, 180; on energy-linked reactions with ATP, 101, 163; on α-ketoglutarate dismutation, 127; on ATP synthesis, 182; on H^+ release, 203; on NTPase, 165; on TPN reduction, 155, 160, 163, 180; inhibitor of Ca^{++} accumulation, 230; of Mg^{++} accumulation, 239; of Mn^{++} accumulation, 230; of NaK ATPase, 249; in ascorbate TMPD system, 52; not as uncoupler, 72

oxidative phosphorylation: general mechanism, 103, 201; nucleotide specificity, 14; restoration of by factors, 84

particles, submitochondrial:
A, 177; T, 177

pH measurements: of ATPase, 128, 195; during ion uptake, 202; release of H^+ by valinomycin, 194

phosphate: effect on energy-linked reactions: 101, 129, 167; effect on pyridine nucleotide translocation, 132; effect on transhydrogenase, 161, 171; in glutamate synthesis, 140; on TPNH oxidation, 147

phosphate exchange, with other triphosphate nucleotides, 163, 164

phosphate potential, effect on transhydrogenase, 154, 176, 177

phospholipids, Mn^{++} binding, 227, 274

photophosphorylation, 208

photosynthesis, bacterial, 207, 216

PMS: DPN reduction with, 7; oxidative phosphorylation with, 24

Pulsed Nuclear Magnetic Resonance (see also spin echo technique), 224; enhancement of relaxation rate, 225-227

rotenone, 160, 179, 221

reactions, energy-linked: 39; DPN reduction, 43; TPN reduction, 45; in chloroplasts, 51; of ubiquinone, 87; definition of, 179

respiration: ATP-activated, 254; Ca^{++} activated, 254;

K_m for Ca^{++} activation, 259

Shoelaces, necessity for, 75

succinate linked reduction of DPN, 39, 97, 98; effect P_i, 101; effect oligomycin, 100; effect antimycin, 98; effect ITP, 163; effect of factors, 83; ATP as energy source, 101; Coupled to transhydrogenase, 159; formation of DPNH$\sim$P, 117; value of n, 102

; n in succinate-linked reduction of pyridine nucleotide, 102, 105, 115; n in aspartate synthesis, 108, 180; n for energy linked transhydrogenase, 160, 168, 176, 179

X: 172, 176

tetramethyl-p-phenylenediamine for cytochrome oxidase phosphorylation, 193; reduction of DPN, 6-13, 23, 99, 107; oxidative phosphorylation with, 24

TPN$\sim$ I, 110

transducer, chemiosmotic, 51

transhydrogenase: effect of ATP, 169; role in energy-linked reactions, 172; general, 109, 122, 143; effect of ATP, 129, 158; effect of ITP, CTP, GTP, UTP, 162, 163, 171; effect of Mg^{++}, 161; in submitochondrial fractions, 177; mechanism, 151, 170, 172, linked to succinate-linked pyridine nucleotide reduction, 159 transport phenomena: general, 122; in pyridine nucleotide compartmentation, 134, 141

thyroxine analogues and triiodothyronine, 180: effect of transhydrogenase, 149, 150, 170; effect of succinate-linked pyridine nucleotide reduction, 150

ubiquinone (see also CoQ): general, 87; role in energy-linked reactions, 99, 106, 123; organization, 122; DPN reduction by, 26

uncouplers: sites of action, 188, 192, 200; specificity, 189

usnic acid, 191

valinomycin: effect on H^+ release, 194, 196